THE EVOLUTION OF ADAPTATION BY NATURAL SELECTION

THE EVOLUTION OF ADAPTATION BY NATURAL SELECTION

A ROYAL SOCIETY DISCUSSION MEETING
ORGANIZED BY

J. MAYNARD SMITH, F.R.S., AND
R. HOLLIDAY, F.R.S.

HELD ON 6 AND 7 DECEMBER 1978

LONDON
THE ROYAL SOCIETY
1979

Printed in Great Britain for the Royal Society
at the
University Press, Cambridge

ISBN 0 85403 118 9

First published in *Proceedings of the Royal Society of London*,
series B, volume 205 (no. 1161), pages 435–604

Published by the Royal Society
6 Carlton House Terrace, London SW1Y 5AG

PREFACE

During the last twenty years, evolutionary biology has been concerned with two main problem areas. The first is to work out the implications for evolution of advances in molecular biology. These implications are profound. In the 'central dogma', that information cannot be transferred from proteins to nucleic acids, we have a biochemical explanation of Weismann's theory of the independence of the 'germ line' from the soma. Once this independence is accepted, Darwin's theory of natural selection becomes true in a stronger sense than he himself was willing to claim. If there is no way in which individually acquired adaptations can be impressed upon the hereditary material, then selection becomes the only source of evolutionary adaptation.

In this symposium, three speakers discuss the molecular basis of evolution. Orgel describes the evolution of adaptation by natural selection in populations of molecules replicating *in vitro*. His paper shows how the study of the origin of life is gradually moving from speculation to theory and experiment. Hartley shows how genetical and biochemical methods can be used to study the evolution of proteins. One importance of this line of work is that it becomes possible to study the origin and evolution of new biochemical adaptations in the laboratory, because of the large population sizes and short generation times of microorganisms. Clarke builds a bridge between molecular biology and classical population genetics – the study of variation in natural populations. Biochemical methods, in particular electrophoresis, have made it possible to measure the immense range of genetic variability in wild populations. This in turn has resurrected the old argument between selection and genetic drift in a new form. How much of the observed genetic variability is selectively neutral?

The second problem area arose from an attempt to explain in selectionist terms a number of characteristics of living things that at first sight seem not to increase the Darwinian fitness of the individuals possessing them. These characteristics – social and apparently 'self-sacrificing' behaviour, sexual reproduction, senescence – raise difficulties that have been discussed for over a century. What is new is a more rigorous attempt to explain their evolution in terms compatible with natural selection and population genetics.

Two events in the early 1960s sparked off this reinvestigation. One was the publication of Wynne-Edwards's *Animal dispersion* in 1962, proposing that social behaviour was in origin an adaptation for regulating animal population numbers, and, since it is a population and not an individual adaptation, that the population and not the individual has been the relevant unit of selection. Such 'group selection' explanations were not new in biology. Indeed, group selection thinking was widespread, if often unconscious. By arguing explicitly that group selection was the explanation of social behaviour, Wynne-Edwards brought this problem into the open. The difficulty lies with the fact that population genetic models suggest that group selection can be effective only in species divided into small and genetically isolated groups.

Wynne-Edwards's explanation of the evolution of social behaviour was challenged by many people, but most notably by G. C. Williams's *Adaptation and natural selection*, published in 1966. At the same time, an alternative mechanism for the evolution of social behaviour had been proposed by W. D. Hamilton in his papers on 'The genetical evolution of social behaviour', which appeared in 1964. The essence of Hamilton's argument is that if one wishes to understand evolution, one must take into account the effects of an individual's actions, not only on its own survival and that of its direct descendants, but also on the survival and fecundity of collateral relatives. Hamilton's theoretical work has led to a new approach to field studies of animal societies, with an emphasis on the genetic relationships between individuals.

A number of papers at this symposium are concerned to find selective explanations of particular phenomena. Maynard Smith uses the methods of evolutionary game theory to interpret contest behaviour in animals. The paper asks how far empirical studies are consistent with theoretical predictions, and identifies phenomena that still require explanation. Dawkins & Krebs extend game theoretic ideas to a dynamic consideration of 'arms races' within and between species. They are particularly concerned with whether such races end in a mutual equilibrium, or in continuous or cyclical evolution.

Kirkwood & Holliday revisit the evolution of senescence, a problem which has repeatedly been discussed, from Weismann to Medawar, Williams and Hamilton. Their paper illustrates both aspects of the symposium, since it is concerned both with selective processes and molecular mechanisms.

Of all current problems of evolution theory, those concerned with the evolution of sex are the most difficult, if only because it is not clear whether sexual reproduction should be seen as an individual or a population adaptation. B. & D. Charlesworth discuss the evolution of sexual systems in flowering plants. Their approach is essentially that of population genetics, the aim being to account for the evolution of selfing and outcrossing mechanisms, and in particular dioecy and distyly, in terms of selection within populations altering the frequencies of genes. Finally, in this group of papers, Williams returns to the evolution of the sex ratio, a topic first presented in its modern guise by R. A. Fisher. Williams points out that selective mechanisms have been proposed which would either increase or decrease the within-progeny variance of the sex ratio. Yet, as he demonstrates, evidence from vertebrates suggests a random and non-adaptive determination of sex – a serious theoretical difficulty.

Perhaps the strongest criticism that can be made of this search for adaptive explanation is that it leads only to speculative adaptive explanations, not easily subject to test. This difficulty is the concern of the papers by Clutton-Brock &

Harvey, and Gould & Lewontin. The latter authors emphasize the difficulty of what they call the 'adaptationist programme'. They emphasize the importance of developmental constraints in limiting the course of evolution. Clutton-Brock and Harvey argue that there are methods whereby adaptationist hypotheses can be tested, and show how a statistical analysis of comparative data can overcome some of the difficulties.

Predicting the future development of science is always a hazardous enterprise, but it is natural to ask how evolutionary biology is likely to develop in the next twenty years. Developments on the molecular aspects of evolution seem certain, because they will be sustained by the progress of molecular biology itself. However, three points are worth making. First, the advantages of microorganisms for the experimental study of evolution must be exploited. Secondly, the study of population structure, genetic variability and selection in natural populations of prokaryotes still awaits its Dobzhansky. Thirdly, the interpretation of prokaryote evolution will have to take into account a whole range of methods of genetic transfer; indeed, a theory in which fitnesses are ascribed to individuals may no longer be appropriate.

In studies of the evolution of higher organisms, the present concentration on behaviour and on the evolution of genetic mechanisms is likely to continue, if only because these are the areas of greatest difficulty for a Darwinian explanation. However, progress in this field depends on our placing less emphasis on the ingenuity and originality of our hypotheses, and more on logical rigour and on empiric testing. In particular, we must learn to treat comparative data with the same respect as we would treat experimental results.

Finally, what of developmental constraints? At least in Britain and America, the significance given to developmental biology in evolution theory has steadily diminished during the past twenty years, and perhaps ever since Weismann's theory made it possible to study genetics without understanding development. We think that this trend is likely to be reversed, but only when new ideas appear within developmental biology that make the concept of a developmental constraint a usable one when studying particular evolutionary events. Until that happens, there will be a guest missing at the feast.

July 1979

J. Maynard Smith
R. Holliday

CONTENTS

Proc. R. Soc. Lond. B **205**, 435–442 (1979)
Printed in Great Britain

Selection *in vitro*

By L. E. Orgel, F.R.S.
The Salk Institute for Biological Studies, P.O. Box 1809,
San Diego, California 92112, *U.S.A.*

The Qβ-polymerase–Qβ-RNA system is reviewed and the evolution of resistance to inhibitors is discussed. It is suggested that this system provides a useful model of the evolution of haploid genomes under natural selection. Consequences for theories of the origins of life are discussed.

1. Introduction

Many writers have suggested that the non-enzymic replication of an RNA- or DNA-like molecule must have been an early and essential step in the origins of life on the Earth. Attempts to demonstrate such a reaction experimentally have not progressed very far (Orgel & Lohrmann 1974), so it would not be useful to review work in this area at a meeting mainly concerned with the evolution of higher organisms. However, there is an enzymic system much less complex than a living organism whose behaviour is likely to throw light on prebiotic and biotic evolution.

The RNA polymerase of the RNA-containing bacteriophage Qβ was first isolated by Haruna & Spiegelman (1965). It is a specific polymerase in the sense that it will bring about the replication of Qβ-RNA but, in general, will not act on unrelated RNA molecules. The enzyme is robust: given a single strand of a suitable RNA substrate, it will bring about the in vitro synthesis of an indefinite number of copies. RNA synthesis is exponential so long as nucleoside triphosphates are available and the enzyme is in excess.

Spiegelman *et al.* (1975) established that, under appropriate conditions, the product made when Qβ-polymerase uses viral RNA as template is active biologically: it can direct the production of normal viruses. On the other hand, the polymerase does occasionally make mistakes, both *in vivo* and *in vitro*. The error rate is probably about 1 in 10^4 under physiological conditions.

These simple properties of Qβ-polymerase and its RNA substrates open up a new domain for the study of evolution under the influence of mutation and natural selection. One no longer needs a living organism to provide the link between the nucleic acid and its expression: one can study natural selection in a test-tube. In particular, protein synthesis, which is involved in one way or another in determining the selective value of most mutations in living organisms, is completely bypassed. Of course, selection now acts at a completely different level: normally, selection 'tests' the appropriateness of a nucleic acid sequence only after the sequence has been 'interpreted' by the protein synthetic apparatus, etc. In our in-vitro system, selection 'tests' directly the physico-chemical properties of an RNA molecule or the interaction between the RNA molecule and its polymerase. The in-vitro system provides a much less rich domain for investigation, but one which is much more easily subjected to analysis at the molecular level.

I should like, first, to make these abstract ideas clear by concentrating on a single example, the development of resistance to the drug ethidium bromide in an in-vitro system. I shall describe how the experiments were done, and then turn to the interpretation of the observed selective values in terms of molecular properties. A great deal of background material must be omitted; it is covered in an extensive review by Spiegelman *et al.* (1975).

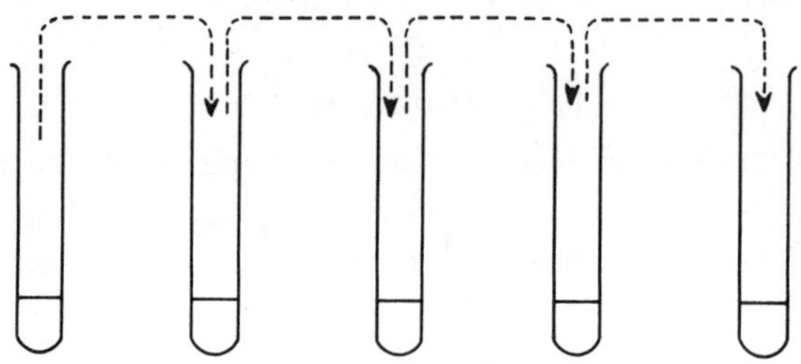

FIGURE 1. Life-cycles (4) of RNA substrates.

2. Resistance to ethidium bromide

In early experiments, Spiegelman and his colleagues established that Qβ-RNA and replicating variants derived from it 'adapt' rather rapidly to unfavourable incubation conditions. They showed, for example, that RNA grown repeatedly in the presence of inhibiting concentrations of the ATP analogue tuberdicin became resistant to the drug, presumably by 'mutating away' those sites where substitution of adenosine by tubercidin is particularly damaging. Similarly, mutants that would grow at very low levels of nucleoside triphosphates were obtained by repeated growth in incubation mixtures containing abnormally low levels of the nucleoside triphosphates.

In our laboratory, the first experiments on the selection of RNAs resistant to ethidium bromide were carried out with a highly purified RNA variant called V2 (Saffhill *et al.* 1970). This RNA is replicated much faster than Qβ RNA by the polymerase: it has only about 550 bases in its sequence compared with the 3300 present in Qβ-RNA. The design of the experiment is illustrated in figure 1.

Each of the first few tubes contains a standard amount of enzyme and incubation mixture and an inhibiting concentration of ethidium bromide (2 μg/ml: 40% inhibition). An aliquot of V2 RNA is introduced into the first tube and allowed to replicate at 38 °C for 30 min. Then 9% (one-eleventh) of the sample from the first tube is added to the second tube and that tube incubated for 30 min. Transfers of this type are repeated until the amount of RNA synthesized per incubation approaches the uninhibited value. At this point a higher concentration of ethidium bromide (4 μg/ml) is added to the incubation mixture in subsequent tubes. The result is shown in figure 2. Synthesis of RNA drops at first but soon recovers (11 transfers). The concentration of drug is then doubled again. By repeating this procedure we obtained, after about 100 transfers, an RNA (V40) which would grow in the presence of 40 μg/ml of drug.

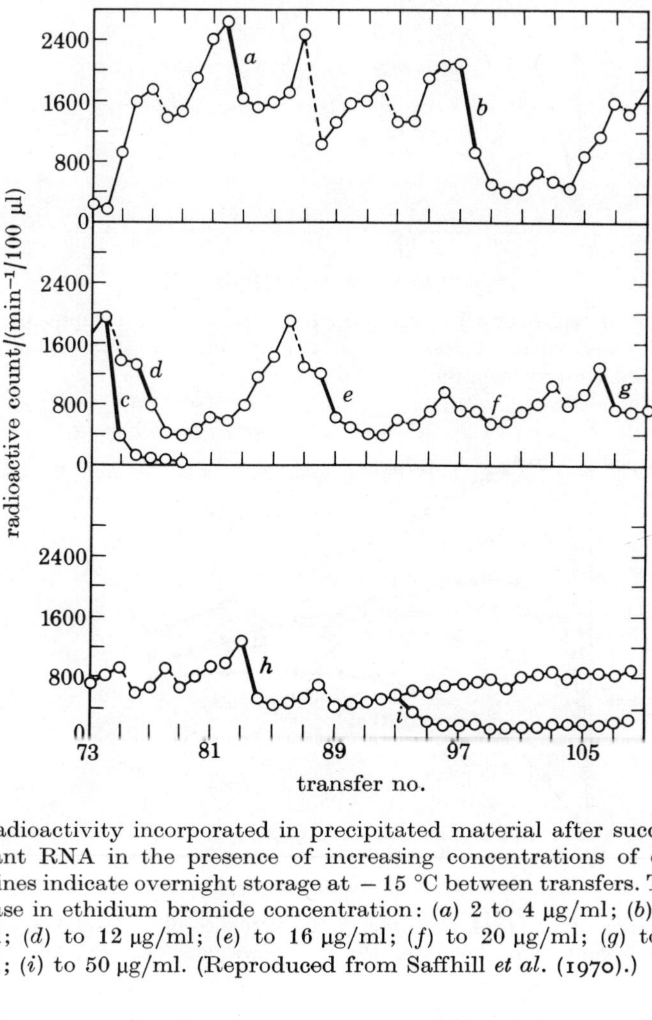

FIGURE 2. Radioactivity incorporated in precipitated material after successive transfers of Qβ variant RNA in the presence of increasing concentrations of ethidium bromide. Broken lines indicate overnight storage at − 15 °C between transfers. Thick lines indicate an increase in ethidium bromide concentration: (*a*) 2 to 4 μg/ml; (*b*) to 8 μg/ml; (*c*) to 16 μg/ml; (*d*) to 12 μg/ml; (*e*) to 16 μg/ml; (*f*) to 20 μg/ml; (*g*) to 30 μg/ml; (*h*) to 40 μg/ml; (*i*) to 50 μg/ml. (Reproduced from Saffhill *et al.* (1970).)

Standard inhibition studies showed that the RNA becomes progressively more resistant to ethidium bromide as the experiment progresses (figure 3). The resistance is not lost rapidly on transfer in the absence of ethidium bromide. Most interestingly, fluorescence studies show that the adapted V40 RNA binds less ethidium bromide than does the starting V2 RNA (figure 4). In the absence of

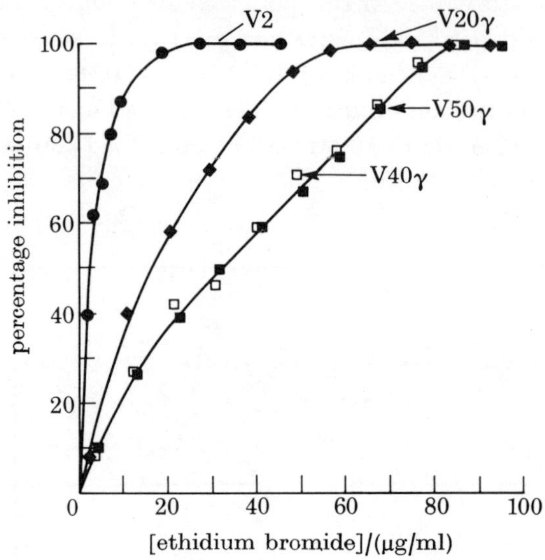

FIGURE 3. Inhibition of replication by ethidium bromide. V2 is the input sequence. V20γ, V40γ and V50γ are variants resistant to 20, 40 and 50 μg/ml ethidium bromide, respectively. (Reproduced from Saffhill *et al.* (1970).)

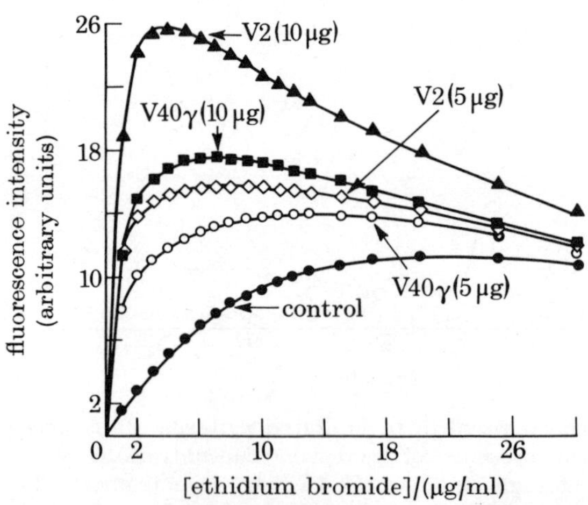

FIGURE 4. Fluorescence of ethidium bromide in the presence of V2 and V40γ in 0.1 M Tris buffer (pH 7.4). The control has no added RNA. (Reproduced from Saffhill *et al.* (1970).)

ethidium bromide, V40 grows at only about 40% of the rate achieved by V2 under the same conditions.

These experiments suggest that V2 RNA molecules bind ethidium bromide strongly at a few special positions in the molecule in such a way as to slow down their replication. Mutant RNAs arise from time to time in which these special sites are modified so that they no longer bind the drug. Selection in the presence of the drug then leads to the outgrowth of resistant mutants. However, the mutants are less well adapted than the wild type to growth in the absence of the drug. These speculations were confirmed in an elegant series of experiments carried out in Spiegelman's laboratory (Kramer *et al.* 1974).

This set of experiments used another RNA variant, MDV-1, which contains only 218 nucleotides in its sequence. The complete sequence of the RNA was determined by Mills *et al.* (figure 5). A series of 25 sequential transfers was carried

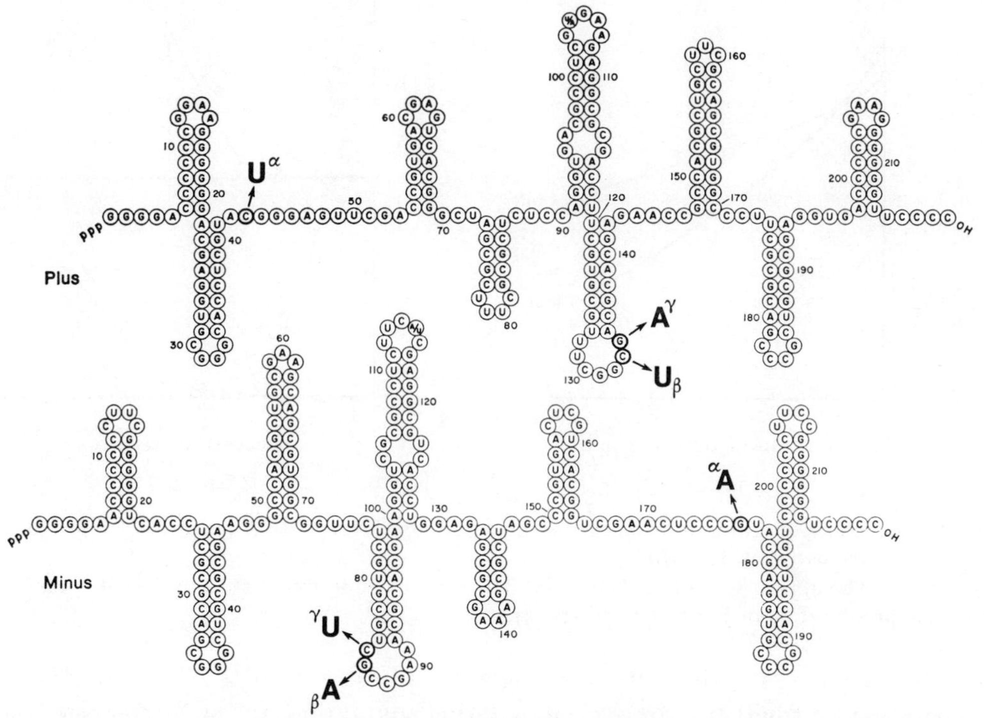

FIGURE 5. Nucleotide sequence of mutant RNA. Both strands of MDV-1 RNA are illustrated. The bold letters and arrows indicate the identity and location of each nucleotide substitution that occurred. The three mutations are arbitrarily labelled α, β and γ. (Reproduced from Kramer *et al.* (1974).)

out in 15 μM ethidium bromide. In these experiments only 10^{-5} of the material in each tube was transferred to initiate the next reaction, so as to maximize the period of exponential growth.

At the end of this series of transfers an RNA population was obtained that was more resistant than the wild-type to ethidium bromide (figure 6). The sequence of this RNA was determined and found to differ from that of MDV-1 at just three positions (figure 5). The amount of drug bound to MDV-1 and to the adapted variant was determined by equilibrium dialysis. The three mutations had reduced drug-binding by about 35 %.

A further series of experiments was carried out to decide whether the final

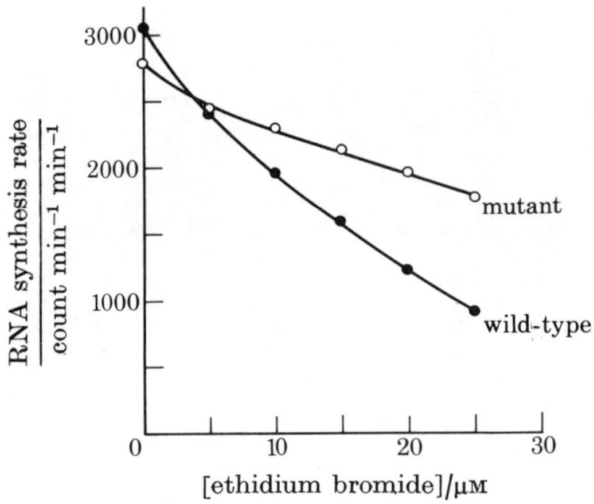

Figure 6

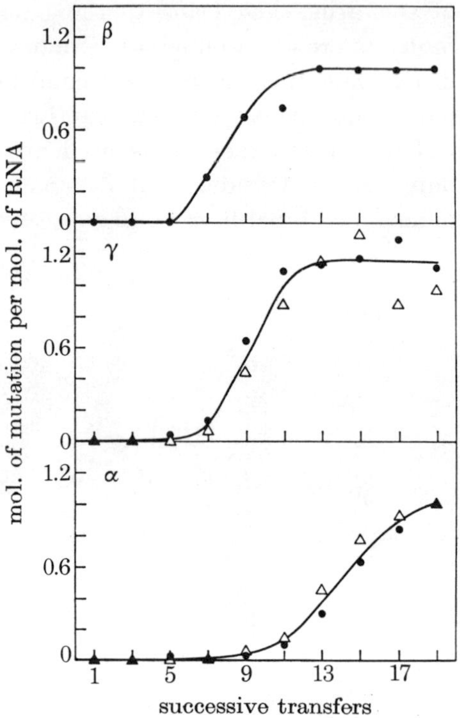

Figure 7

Figure 6. Effect of ethidium bromide on mutant and wild-type RNA synthesis. (Reproduced from Kramer *et al.* (1974).)

Figure 7. The appearance of mutations α, β and γ as a function of the number of transfers. (Reproduced from Kramer *et al.* (1974).)

adapted RNA was derived from a single variant strand in the original MDV-1 preparation, or whether it arose *de novo* during adaptation. In the former case, the three mutations α, β and γ should have been present in equal amounts at each stage of adaptation; in the latter, the mutations would have appeared at different stages in the transfer series. The results shown in figure 7 indicate clearly that the mutation α appeared after β and γ showing that the resistant variant arose *de novo* A more detailed analysis showed that β appeared before γ.

3. Evolution of RNAs

The ease with which we can select RNA molecules that have lost some of their binding sites for ethidium bromide is highly suggestive. Many important control mechanisms involve the interaction of nucleic acids with proteins, for example the interaction of DNA with RNA polymerase and the interaction of RNA with ribosomal proteins. The alteration of the sequence of an RNA so as to change its interaction with a protein should be achieved at least as easily as the alterations which change the RNA–ethidium bromide interaction. As a result of our work with ethidium bromide, we concluded that many aspects of cellular control are likely to evolve through the modification of the sequences of promotors, operators, etc., more readily than by the modification of the proteins with which they interact.

The direct interaction of polynucleotides with small molecules is likely to have been particularly important at the earliest stages in the evolution of life. The origin of the genetic code, for example, may have depended on the stereospecific interaction of amino acids with tRNA-like molecules. Unfortunately, virtually nothing is known about the strength and specificity of these interactions, and the subject is not easily attacked by conventional methods. The selective experiments with ethidium bromide, although only indirectly relevant, strongly suggest that specific positive interactions could evolve relatively quickly, at least in the case of amino acids such as tryptophan and tyrosine which have large, planar side chains. I suspect that a fair level of discrimination could be achieved between other pairs of amino acids.

Finally, I should like to emphasize the wide range of experiments in natural selection that could be done with the use of the Qβ system. It should be trivial for example to select for molecules that (1) stick (or fail to stick) to a chosen surface; (2) pass through (or fail to pass through) a porous barrier; (3) are resistant to destruction by an enzyme or other physical or chemical agent.

With slightly greater ingenuity it should be possible to select for molecules that 'recognize' small molecules in their environment and utilize such molecules to increase their own lifetime or rate of replication or even to interfere with the replication of their competitors. Here perhaps lies an approach to the origin of the 'selfish gene'. I find it surprising that so little has been done along these lines; perhaps it is because experiments in this field are difficult to fund.

The Qβ polymerase system has been studied more extensively than any other in the context of natural selection *in vitro*. However, it is not unique. We found that the DNA-dependent polymerase of *Escherichia coli*, under appropriate conditions, will select from a totally random mixture of polynucleotides one or more strands which it is capable of replicating indefinitely. This suggests (Biebricher & Orgel 1973) that any RNA polymerase will find some substrates on which it acts as a replicase, even if the enzyme does not have a replicase function in nature.

4. Population structure

The theory of natural selection is concerned with the development of populations in time. At first sight, this does not seem to be so for the Qβ system *in vivo* or *in vitro*: the wild-type sequence appears on the printed page as a unique sequence of bases. However, a more detailed consideration of the experimental situation shows that the idea of a unique, wild-type sequence is an oversimplification. A typical analysis of an RNA would surely pick up ambiguity at a position in the sequence if the two alternative bases at that position were present in comparable amounts, but would probably miss ambiguity below the 10% level and would certainly miss it at the 1% level. A great deal of variation could, therefore, be present without any being detected by the normal procedures of sequence analysis. Eigen has emphasized that a wild-type population might contain only a very small fraction of wild-type sequences (Eigen & Schuster 1977).

The mutation rate for an RNA polymerase is typically 10^{-4} per base per replication. Suppose that the selective disadvantage of a typical mutation is 10%. Then, at equilibrium, one must expect a mutant frequency of about 10^{-3} at a typical site. The majority of Qβ phages in a wild type population would have at least 1 mutation among their *ca.* 3000 bases. Under in-vitro conditions mutation rates are likely to be higher and selection pressure may well be much weaker. Thus the amount of variation could be much larger. Population structure of this kind will certainly prove important for the analysis of the rate of evolution of RNA populations.

References (Orgel)

Biebricher, C. K. & Orgel, L. E. 1973 *Proc. natn. Acad. Sci. U.S.A.* **70**, 934.
Eigen, M. & Schuster, P. 1977 *Naturwissenschaften* **64**, 541.
Haruna, I. & Spiegelman, S. 1965 *Proc. natn. Acad. Sci. U.S.A.* **54**, 579.
Kramer, F. R., Mills, D. R., Cole, P. E., Nishihara, T. & Spiegelman, S. 1974 *J. molec. Biol.* **89**, 719.
Orgel, L. E. & Lohrmann, R. 1974 *Acct. chem. Res.* **7**, 368.
Saffhill, R., Schneider-Bernloehr, H., Orgel, L. E. & Spiegelman, S. 1970 *J. molec. Biol.* **51**, 531.
Spiegelman, S., Mills, D. R. & Kramer, F. R. 1975 In *Stability and origins of biological information* (ed. I. R. Miller), p. 183. New York: Halsted Press.

Proc. R. Soc. Lond. B **205**, 443–452 (1979)
Printed in Great Britain

Evolution of enzyme structure

BY B. S. HARTLEY, F.R.S.
Department of Biochemistry, Imperial College of Science and Technology, London SW7 2AZ, U.K.

Three-dimensional structures of enzymes offer evidence about their evolution. There are clear examples of divergent families (e.g. mammalian serine proteases) and convergence (e.g. chymotrypsin and subtilisin). Topological similarities in dehydrogenases may reflect an ancient divergence or merely chemical constraints on protein architectures.

Further experimental evidence is desirable to back up arguments based on molecular morphology. By growing microorganisms on novel foodstuffs in a chemostat, one can focus selective pressure on a specific enzyme activity. Experiments will be described in which such pressure is focused on pentitol metabolism. Examination of the fine structure of the genes responsible for this pentitol metabolism has given clues about the evolution of metabolic pathways.

IS MOLECULAR GENETICS NOW A BRANCH OF CHEMISTRY?

Over the last 25 years, biochemistry has made a major impact on evolutionary biology. In 1953 we knew that genes must somehow be encoded in the structure of DNA (Avery *et al.* 1944) and Sanger & Thompson (1953) had just published the structure of bovine insulin, proving that proteins had a covalent structure of 20 different amino acids linked by α-peptide bonds. But the essential functions of a gene – to replicate exactly and to specify exactly the synthesis of a protein such as an enzyme with a particular biology activity – were shrouded in mystery.

Although today we lack many details of the processes of gene replication and expression, every step is to a large degree explicable in chemical terms. Hence this quarter of a century has seen the subject of genetics evolve from its long history as a branch of biology to what some might now consider as a branch of chemistry.

For example, in 1977 Sanger *et al.* described the complete covalent structure of the genome of a living organism: the bacterial virus ϕX 174. Knowing the genetic code and a series of plausible rules for beginning and ending regions of translation, they were able to predict the complete amino acid sequences of the proteins that could be encoded in this genome. To everyone's surprise it emerged that the hitherto accepted dogma of a triplet, non-overlapping genetic code was untrue: a single stretch of nucleotides could code for two and possibly even three different proteins, depending on the phase of initiation and translation. Moreover, a sequence that was already coding for two proteins might also code at the level of DNA for a promoter site for initiating transcription, or at the level of RNA for a site for binding ribosomes thereby controlling rates of translation. In this dense concentration of biological information it is difficult to define that simple concept of the gene that has served biologists so well in the earlier part of this century. Words such as cistrons, replicons, operons and more recently, introns and exons, have been coined to amplify and amend the earlier concept of one gene – one protein – one biological activity.

However, the chemist simply sees ϕX 174 as a sequence of 5375 nucleotides that he and the *Escherichia coli* cell can translate into several different proteins. If he knew the rules of folding for a polypeptide chain – and he is confident that this is a problem that a good chemist will shortly resolve – he could build three-dimensional structures of these proteins. With even more optimism and a bit of luck thrown in, he might guess the enzymic activity and specificity of some of these proteins and thereby predict their biological role. Hence he might view molecular genetics as a branch of chemistry.

Species differences in amino acid sequences

How has this new outlook affected evolutionary biology? Going back to 1953, having sequenced bovine insulin Sanger began to study the sequences of insulin from sheep and pigs. I recall arguing with him at the time that this was a relatively trivial problem since we already knew from the work of Du Vigneaud *et al.* (1953) that the nine-residue peptide vasopressin differed by a change of Lys-8 in pigs to Arg-8 in cows. One might predict similar changes in a few residues in the sheep and pig insulins, but so what? In fact, Sanger and his colleagues (Brown *et al.* 1955) discovered that the only changes were in residues 9, 10 and 11 in the A-chain of the insulins: –Ala–Ser–Val– in cows, –Ala–Gly–Val– in sheep and –Thr–Ser–Ile– in pigs. One could construct a sort of evolutionary tree in which a cow with –Ala–Ser–Val– evolved into a sheep with –Ala–Gly–Val– or by two steps into a pig with –Thr–Ser–Ile–. Not a very plausible tree, obviously, but might more sequences from more species produce trees more satisfying to the classical biologists?

In fact Sanger's initiative generated a considerable industry in comparative protein chemistry, particularly with cytochromes, haemoglobins and fibrinopeptides, with some implications that have disturbed classical phylogenies; Zuckerkandl & Pauling (1962) observed that the haemoglobin sequences appeared to have evolved at constant rates, constant that is with respect to time rather than number of generations. On the whole this observation appears to have stood the test of time, although the degree of error in calibrating times of species divergence with palaeontological evidence represents a serious defect in this 'evolutionary clock' hypothesis (Wilson *et al.* 1977). From this concept of the 'molecular clock' arose the 'neutral drift' hypothesis of Kimura (1968) that appears to offer an alternative to the Darwinian approach to the origin of sequence diversity in proteins. This assumes that most of the sequence diversity is not due to natural selection, but is caused by random fixation of selectively neutral amino acid replacements controlled only by mutation rates.

Another observation arising from sequence comparisons of the same protein in different species was that different proteins appeared to have evolved at different rates. If we estimate such rates as the time required for two lineages to diverge by 1 % from a putative common ancestor, histones would appear to have evolved 100 times slower than fibrinopeptides (Wilson *et al.* 1977).

Faced with the mass of information that is fed into these sequence comparisons and the increasing sophistication of the mathematical analysis that is allowed by computer programming, it is important that the average biologist appreciates the defects of some of the basic assumptions that lie behind the mathematics. The raw data consist of percentage sequence variabilities of proteins with similar biological activity extracted from different living species. One then assumes a series of ancestral sequences that minimize the number of sequence changes in each branching lineage, giving an evolutionary tree. Knowing the genetic code, one can improve the data by translating back to hypothetical DNA sequences of the ancestral genes and minimizing the number of point mutations necessary to achieve the observed sequence changes. One can also attempt to allow for potentially 'silent' point mutations in the third base of the codon and for reversions in which sequential point mutations would restore the original amino acid.

One must admit that in most cases evolutionary trees generated in this way bear a comforting resemblance to classical phylogenies. Recall, however, the common basic assumptions: all sequence changes arise by a pathway involving a minimum number of point mutations and all positions in a sequence are treated as equally mutable. It is immediately obvious from inspection of the sequence data that such assumptions must be excessively naïve. Changes at a particular position range from invariant through chemically conservative to hypervariable. The number of changes that would require two point mutations is in excess of statistical expectation and deletions are fairly common.

The reasons for these discrepancies become apparent when we look at the three-dimensional structure of the protein. Amino acid side chains that are internal can obviously vary only within narrow limits without affecting the architecture of the polypeptide chain that is clearly paramount in determining the biological activity. Similarly, residues lying within the substrate binding site of an enzyme must be conserved to retain a common catalytic activity and specificity.

This still leaves a large number of side chains at the surface of a globular protein that could vary without affecting activity, and indeed it is here that species differences are most abundant. A protein chemist could be forgiven for accepting such species variation as 'neutral'. Such arguments were reasonably adduced for cytochrome *c* since a large part of the surface of this small protein, whose role is to conduct electrons from cytochrome *b* to cytochrome oxidase, is very variable between species. Yet all species of cytochrome *c* appeared to react equally well *in vitro* with a preparation of cytochrome oxidase from horse heart. However, doubt has been cast recently on this dogma by Margoliash, himself a proponent of the original hypothesis, by the observation that the in-vitro assay measures properties of the cytochromes *c* under conditions that are far from physiological.

Assays carried out under more physiological conditions reveal a marked specificity of a cytochrome oxidase for cytochrome *c* of its own species (Ferguson-Miller *et al.* 1976). Hence the temptation must be avoided to view amino acid replacements in tertiary structure of proteins as 'neutral' simply because one cannot obviously discern their selective value. Selection must also operate on residues that determine the intracellular location of the protein, its interaction with its environment (e.g. with ions or other proteins) or its stability. Indeed one would expect hypervariability in just those residues where such fine adjustments to small changes in environment could be made without prejudice to the essential biological activity. Moreover, our knowledge of DNA and RNA structure makes it increasingly likely that selective constraints at these levels of gene expression will limit the variability of amino acid sequence even at a putatively 'neutral' site in the protein structure. Hence the concept that 'neutral mutations' can account for a large proportion of sequence variations in proteins looks increasingly unattractive to the sophisticated protein chemist.

The observation that different proteins appear to evolve at different rates also appears to be no problem if we consider tertiary structures of proteins. Internal residues will be strongly conserved for reasons of architecture, while surface residues involved in binding substrates, other protein subunits or membrane components will also be strongly conserved. Since the proportion of these will vary widely, wide divergence in the percentage of replacements in different proteins in a particular lineage is to be expected rather than remarked upon.

Where do new enzymes come from?

Evolution is obviously not limited to reshuffling a pack of primeval genes. New enzyme activities have clearly arisen in response to selective pressures, and the total number of different enzymes in some organisms has increased markedly. Light has been cast on this process by studying the structures of enzyme families: groups of enzymes with common features of catalytic activity and mechanism that can differ widely in substrate specificity.

I have argued elsewhere (Hartley 1966, 1974) that the evolution of such a family from a common ancestor implies a first step of gene duplication, elicited probably by a demand for more of the enzyme. If this selective constraint were then removed, mutations that led to an inactive product in one copy of the gene would be conserved in the population by virtue of the activity of the product of its unchanged partner. Thenceforth the original gene would only tolerate conservative sequence changes, but the 'silent gene' would be free to accumulate all mutations, since it is effectively released from selective pressure. However, from this highly variable 'silent gene' pool, a mutation that repaired the ability of the protein to refold would restore an enzymic capacity together with 'neutral mutations' accumulated during the silent phase. The substrate specificity might then be similar to its original ancestor, yielding isoenzymes with similar specificity but differing in sequence. However, this hypothetical silent gene pool would also be a rich source of enzymes with new specificities. Clearly selection would then favour refinement of these structures to their new biological roles, but the structures might still bear the imprints of stochastic events garnered in their silent phase of evolution.

Structures of serine proteases

The serine proteases, for example, represent such an enzyme family. In mammalian pancreatic juice there are zymogens (inactive precursors) for five such proteases: chymotrypsins A, B and C with a specificity for aromatic side chains, trypsin with a specificity for basic side chains, and elastase, which cleaves adjacent to small hydrophobic side chains.

The detailed tertiary structures of chymotrypsin A, trypsin and elastase show astonishing similarity in the conformation of their polypeptide chains. Apart from deletions or insertions of amino acids that occur at surface loops of the antiparallel β sheets that compose most of the secondary structure, the polypeptide backbones of the three enzymes are superimposable within 1–2 Å (0.1–0.2 nm). This is achieved despite differences of about 60 % between their amino acid sequences and despite the fact that they have been selected for radically different substrate specificities. In fact the specificity differences result from only one or two changes in sequence: Asp-189 in trypsin for Ser-189 in chymotrypsin, or Val-216 and Thr-226 in elastase for Gly-216 and Gly-226 in chymotrypsin.

Less well understood, but probably even more interesting, is a large family of serine proteases in plasma that is involved in the complex processes of blood clotting, complement fixation and kinin formation. These enzymes seem to be remarkably similar to trypsin except that they have high specificity for particular stretches of amino acid sequence. Thrombin, for example, rapidly splits fibrinopeptides A and B from the N-terminus of two of the chains of fibrinogen but has negligible activity against most other proteins even when they are denatured. Similarly Factor X*a* activates prothrombin by splitting –Ile–Glu–Gly–Arg– sequences in two different parts of the chain, but shows little activity towards other sequences.

The concept of these blood clotting enzymes as a family is strengthened by comparison of the primary structures of their inactive precursors, Factor IX, Factor X and prothrombin, which act at the end of the proteolytic activation cascade. These zymogens differ from the pancreatic zymogens in that large 'domains' of N-terminal sequence precede the C-terminal 200–250 residues that are homologous with trypsin. Prothrombin, for example, has an N-terminal 'Gla domain' which contains residues of γ-carboxyglutamic acid (Gla), formed by a vitamin K dependent carboxylation during synthesis of prothrombin in the liver. This is followed by two disulphide-bridged domains (named 'Kringle-domains' by Magnusson after the traditional pattern on Danish cakes) which are homologous with each other. Then follows the A-chain and finally the B-chain of thrombin, the latter being homologous with trypsin (Magnusson *et al.* 1975). Factor X consists of a 'Gla-domain' and a single 'Kringle-domain', similar to those of prothrombin, followed by the C-terminal 'trypsin domain'. Factor IX appears to have a 'Gla domain' and a 'trypsin domain' but no 'Kringle domain' (Claeys *et al.* 1976). These structures point to an evolution in which genes for a serine protease have duplicated and fused with genes for proteins that are probably involved in binding to micelles of degraded cell membranes. Divergent evolution seems to be the rule in the origin of this large family of mammalian proteases.

We also know a good deal about the structures of several serine proteases from microorganisms. There seem to be three classes of tertiary structure. *Steptomyces griseus* is a rich source of these enzymes and appears to make examples of each class. *S. griseus* trypsin is homologous to bovine trypsin in sequence (34 % identity) and model-building indicates that they probably have similar tertiary structures (Jurášek *et al.* 1976). However, it lacks the basic surface residues that are responsible for binding the acidic activation peptide; hence it may not exist as a zymogen.

S. griseus also makes two other proteases, the A-protease and the B-protease, with a rather broad substrate specificity, that show high homology with each other and also with an elastase-like enzyme named α-lytic protease from *Myxobacter* 495 (now classified as *Lysobacter enzymogenes*). However, these three bacterial enzymes show only 20 % sequence homology with the mammalian enzymes and these are located in very limited areas of the sequence. McLachlan & Shotton (1971) pointed out that the large insertions or deletions necessary to bring these homologies into phase with the mammalian sequences indicate a structure in which only the central core of these enzymes can have structural homology with chymotrypsin.

The tertiary structures have been determined of the B-protease (Delbaere *et al.* 1975) and the A-protease (Brayer *et al.* 1978) of *S. griseus* and of the α-lytic protease of *L. enzymogenes* (M. N. G. James, personal communication). These three structures are similar and about 60 % of the chains show topological equivalence to chymotrypsin, despite the limited sequence homology. This does indeed represent the core of the enzyme and includes residues forming the catalytic sites, which are identical in all these enzymes.

In contrast, there are distinct differences between these three microbial enzymes and the mammalian class. Topologically different external loops are wrapped around the common internal core and an insertion of two residues (192A and 192B) eliminates the substrate binding pocket that is a feature of the mammalian enzymes. Moreover, the N-terminus, although homologous in sequence with the mammalian class, is located in a different part of the three-dimensional structure. In the mammalian enzymes this sequence is involved in activation of the zymogen; the bacterial enzymes seem not to need zymogen precursors since they are secreted directly into the extracellular medium.

Finally, *S. griseus* appears to secrete another class of serine protease homologous with subtilisin of *B. subtilis* (Awad *et al.* 1976). Subtilisin has a quite different tertiary structure from all of the above serine proteases, based on a packing of α-helices rather than β-sheets, yet the catalytic and substrate-binding residues are identical in structure and conformation to the chymotrypsin class although grafted on to quite different elements of secondary and tertiary structure (Kraut *et al.* 1971). Subtilisin and chymotrypsin represent a clear example of convergent evolution at the molecular level.

Evolutionary lessons from the serine proteases

Remarkable conservation of tertiary structure is the outstanding lesson that we learn from this divergent enzyme family. Differences in specificity are readily grafted on to a common specificity pocket in the pancreatic enzymes by only one or two amino acid changes, and we see isoenzymes like chymotrypsins A, B and C with considerable surface sequence differences but identical substrate specificities.

The blood clotting enzymes appear to have evolved zymogens by fusing a C-terminal trypsin-like gene with genes for proteins with the property of binding to phospholipids and of forming multienzyme aggregates at this membrane surface.

S. griseus makes an enzyme remarkably similar to bovine trypsin; this implies identical backbone conformations throughout evolution from a common ancestor. It also makes another class of serine protease that has a chymotrypsin-like core with topologically different loops of chain. This implies fusion of segments of the chymotrypsin structure with quite different elements of another structural gene. Finally, *S. griseus* makes an enzyme-like subtilisin – a serine protease that shows convergent evolution to a common catalytic mechanism.

How could all these three structures evolve independently in some single ancestor of *S. griseus*? Having found one solution to making a decent serine protease, surely it would be simpler to multiply genes and diverge specificities, as we see in mammals? I cannot avoid the conclusion that these structures evolved independently in three different species and were assembled in *S. griseus* by interspecies transfer, perhaps virally mediated – a possibility that I referred to some years ago by suggesting that the bacterium might have been infected by a cow! (Hartley 1970).

I also retain my earlier belief that it is difficult to see how these divergent structures could evolve from a common ancestor by a simple process of sequential point mutations without postulating genetic intermediates that would produce proteins incapable of folding correctly (Hartley 1974). One would now go further and suggest that some radical reshuffling of fragments of a serine protease gene is involved in the divergence of the *S. griseus* proteinase-A class and the chymotrypsin class. If this is so, then 'silent gene' intermediates are surely implied, and my above agrument that these could lead to accumulation of 'neutral mutations' would apply.

Experimental enzyme evolution

Bacterial populations adapt so rapidly to defined selective pressures that one can hope to set up experimental systems that will mimic natural evolutionary events. Our own approach to this problem has been to study the growth of *Klebsiella aerogenes* in chemostats with xylitol as limiting carbon source. Under these conditions the activity of ribitol dehydrogenase becomes growth-limiting, and mutants are selected after spontaneous or induced mutagenesis that have either increased levels of the original enzyme (superproducers) or mutant enzymes with improved activity towards xylitol (specificity mutants) (Rigby *et al.* 1974; Hartley *et al.* 1976).

In our earlier experiments (Rigby *et al.* 1974), superproducers predominated and we concluded from analyses of the frequencies of reversion to low production or mutation to inactive enzyme, that tandem gene duplications were a frequent cause of enzyme superproduction, which can amount to up to 30% of the soluble protein in the cell. In order to analyse such events more rigorously, we transferred the genes for pentitol metabolism from *K. aerogenes* into the chromosome of *E. coli* K12 (Rigby *et al.* 1976). Charnetzky & Mortlock (1974) established that the order of these genes in *K. aerogenes* is:

rbt K – *rbt D* – *rbt C* – *dal B* – *rbt B* – *dal C* – *dal D* – *dal K*,

where *rbt D* codes for ribitol dehydrogenase which converts ribitol to D-ribulose and *rbt K* for ribulokinase which gives D-ribulose-5-phosphate by using ATP. The metabolism of D-arabitol by D-arabitol dehydrogenase (the product of *dal D*) is similar to give D-xylulose followed by xylulokinase (encoded by *dal K*) to give D-xylulose 5-phosphate. *rbt B*, *rbt C* and *dal B* and *dal C* are control genes for the two operons, which are induced by D-ribulose and D-arabitol respectively. Growth on xylitol is via a weak side specificity of a constitutive ribitol dehydrogenase followed by xylulokinase (which is induced by back conversion of some of the xylulose to D-arabitol) (Wu *et al.* 1968).

The *E. coli* K12 – *K. aerogenes* hybrids also evolved like *K. aerogenes* when grown on xylitol in chemostats; i.e. either superproducers of ribitol dehydrogenase or specificity mutants were rapidly selected (Hartley *et al.* 1976). Almost all of the mutant ribitol dehydrogenases appeared to be different, either in kinetic parameters towards ribitol, xylitol or L-arabitol (another poor substrate), temperature stability or electrophoretic mobility. Hence there seem to be many different point mutations that can increase the specificity of the enzyme for xylitol.

Most of the superproducers elicited by growth of these *E. coli* – *K. aerogenes* hybrids on xylitol again showed high frequencies of spontaneous reversion to low enzyme production, indicative of tandem gene duplication. One can conclude that gene duplication is readily elicited by selection for more of an enzyme whose rate is growth-limiting. This is the first step in the hypothesis of how enzyme families evolve (Hartley 1966).

The second step would be mutation of one of the two genes to give an inactive product, followed by accumulation of mutations in this now 'silent' gene. We attempted to test this hypothesis by growing a gene-doubled strain of *K. aerogenes* for several generations on a substrate other than xylitol, e.g. ribitol, glucose or inositol (Hartley *et al.* 1976). On switching back to growth on xylitol we hoped that specificity mutants might readily arise. However, we found only selection of yet higher superproducers, probably because gene triplication can arise readily from tandem identical genes by unequal crossover during replication. During growth on inostiol, in contrast, there was rapid selection of the single *rbt D* revertant. Catabolite repression reduces the synthesis of ribitol dehydrogenase during growth on glucose or ribitol. Inositol does not cause catabolite repression so the superproducing strain will make large amounts of an unnecessary enzyme when grown on this substrate. It has been debated (Koch 1972) whether this would be a serious selective disadvantage; these results indicate that it is.

Fine structure of the pentitol operons

Since I have stressed the role of gene duplication in the origin of enzyme families, it is obvious that we should like to know more about the gene duplications that we have selected experimentally. Moreover, it will not have escaped the reader's attention that the ribitol and D-arabitol operons themselves invite speculation about a possible common ancestor. Having moved these genes from *K. aerogenes* into *E. coli* K12, it became possible to study their fine structure.

To this end, we constructed specialized transducing phages, λp *rbt*, which contains the *rbt* operon or λp *rbt dal*, which contains the *rbt* and *dal* operons inserted into the genome of bacteriophage λ (Neuberger & Hartley 1979). By restriction endonuclease mapping, we analysed the physical structure of these two phages and also located the approximate positions of transcription of the *rbt* and *dal* operons. By separating the two strands of λ DNA and hybridizing these with either *rbt*-mRNA from a superproducing strain of *K. aerogenes*, or *dal*-mRNA formed after induction of the *dal* operon by D-arabitol, we showed that the operons are transcribed in bipolar fashion from opposite strands of the DNA. This suggests that invert gene duplication might have lain on the pathway of evolution of these operons from a common ancestor. Invert gene duplication would avoid the problem of loss of a duplicated copy during recombination, such as we observed experimentally in the example quoted above. This would allow time for the the two sequences to diverge to new enzymes.

Finally, we have begun to analyse the fine structure of the postulated duplications of the *rbt D* gene in our family of strains superproducing ribitol dehydrogenase. In one case, *K. aerogenes* strain A3, we have found a small tandem duplication that embraces the *rbt D* and *rbt K* genes but not *dal D* or *dal K*. In other cases the duplication is probably larger and is not so far revealed by our analysis. We hope to investigate the DNA sequences involved in these events, so as to determine what actually happens when a gene evolves in response to a defined selective pressure.

References (Hartley)

Avery, O. T., McLeod, C. M. & McCarty, M. 1944 *J. exp. Med.* **79**, 137–158.

Awad, W. M., Brew, K., Powell, J. T., Russin, D. J., Seber, J. F., Siegel, S. & Vosbeck, K. D. 1976 In *Proteolysis and physiological regulation* (Miami Winter Symposia) (ed. D. W. Ribbons & K. D. Brew), vol. 2, pp. 77–91. New York: Academic Press.

Brayer, G. D., Delbaere, L. T. J. & James, M. N. G. 1978 *J. molec. Biol.* **124**, 243–283.

Brown, H., Sanger, F. & Kitai, R. 1955 *Biochem. J.* **60**, 556–565.

Charnetzky, W. T. & Mortlock, R. P. 1974 *J. Bact.* **119**, 176–182.

Claeys, H., Sottrup-Jensen, L., Zajdel, M., Petersen, T. E. & Magnusson, S. 1976 *FEBS Lett.* **61**, 20–24.

Delbaere, L. T. J., Hutcheon, W. L. B., James, M. N. G. & Thiessen, W. E. 1975 *Nature, Lond.* **257**, 758–763.

Du Vigneaud, V., Lawler, H. C. & Popenoe, E. A. 1953 *J. Am. chem. Soc.* **75**, 4880–4881.

Ferguson-Miller, S., Brautigan, D. L. & Margoliash, E. 1976 *J. biol. Chem.* **251**, 1104–1115.

Hartley, B. S. 1966 *Advmt Sci., Lond.*, May, pp. 47–54.

Hartley, B. S. 1970 *Phil. Trans. R. Soc. Lond.* B **257**, 77–87.

Hartley, B. S. 1974 In *Evolution in the microbial world* (*Soc. Gen. Microbiol. Symp.* no. 24) (ed. M. J. Carlile & J. J. Skehel), pp. 151–182. Cambridge University Press.

Hartley, B. S., Altosaar, I., Dothie, J. M. & Neuberger, M. S. 1976 In *Structure–function relationships of proteins* (3rd John Innes Symp.) (ed. R. Markham & R. W. Horne), pp. 191–200. Amsterdam: Elsevier/North Holland.

Jurášek, L., Olafson, R. W., Johnson, P. & Smillie, L. B. 1976 In *Proteolysis and physiological regulation* (Miami Winter Symposia) (ed. D. W. Ribbons & K. Brew), vol. 2, pp. 93–123. New York: Academic Press.

Kimura, M. 1968 *Nature, Lond.* **217**, 624–626.

Koch, A. L. 1972 *Genetics, Princeton* **72**, 297–316.

Kraut, J., Robertus, J. D., Birktoft, J. J., Alden, R. A., Wilcox, P. E. & Powers, J. C. 1971 *Cold Spring Harb. Symp. quant. Biol.* **36**, 117–123.

Magnusson, S., Petersen, T. E., Sottrup-Jensen, L. & Claeys, H. 1975 In *Proteases and biological control* (ed. E. Reich, D. B. Rifkin & E. Shaw), pp. 123–149. Cold Spring Harbor Laboratory.

McLachlan, A. D. & Shotton, D. M. 1971 *Nature, new Biol.* **229**, 202–205.

Neuberger, M. S. & Hartley, B. S. 1979 *J. molec. Biol.* (submitted).

Rigby, P. W. J., Burleigh, B. D. & Hartley, B. S. 1974 *Nature, Lond.* **251**, 200–204.

Rigby, P. W. J., Gething, M. J. & Hartley, B. S. 1976 *J. Bact.* **125**, 728–738.

Sanger, F., Air, G. M., Barrell, B. G., Brown, N. L., Coulson, A. R., Fiddes, J. C., Hutchison, C. A., Slocombe, P. M. & Smith, M. 1977 *Nature, Lond.* **265**, 687–695.

Sanger, F. & Thompson, E. O. P. 1953 *Biochem. J.* **53**, 366–374.

Wilson, A. C., Carlson, S. S. & White, T. J. 1977 *A. Rev. Biochem.* **46**, 573–639.

Wu, T. T., Lin, E. C. C. & Tanaka, S. 1968 *J. Bact.* **96**, 447–456.

Zuckerkandl, E. & Pauling, L. 1962 In *Horizons in biochemistry* (ed. M. Kasha & B. Pullman), pp. 189–225. New York: Academic Press.

Proc. R. Soc. Lond. B **205**, 453–474 (1979)
Printed in Great Britain

The evolution of genetic diversity

By B. C. Clarke
Genetics Research Unit, Queen's Medical Centre,
Clifton Boulevard, Nottingham NG7 2UH, U.K.

The existence within natural populations of large amounts of genetic variation in molecules and morphology presents an evolutionary problem. The 'neutralist' solution to this problem, that the variation is usually unimportant to the organisms displaying it, has now lost much of its strength. Interpretations that assume widespread heterozygous advantage also face serious difficulties. A resolution is possible in terms of frequency-dependent selection by predators, parasites and competitors. The evidence for pervasive frequency-dependent selection is now very strong. It appears to follow naturally from the behaviour of predators, from the evolutionary lability of parasites, from the ecology of competition and, at the molecular level, from the phenomena of enzyme kinetics. Such selection can explain the maintenance not only of conventional polymorphism but also of continuous variation in both molecular and morphological characters. It can account for the occurrence of diversity within groups of haploid and self-fertilizing organisms, and for the evolution of differences between individuals in their systems of genetic control.

1. Introduction

Nature is often complicated, but we come to understand it by making and testing simple hypotheses. Because any set of phenomena can be explained by an infinite number of hypotheses, we must somehow choose between them. In practice, the value of a hypothesis is judged by its generality, its suitability for attempts at refutation, and its simplicity. The last criterion is needed because the first two do not sufficiently limit the number of alternatives. Although the simplest explanation may not always be the correct one, the scientific method requires us to accept it as a working hypothesis until the facts demand its rejection. Operationally, we should always prefer the simple to the complex, and the singular to the plural.

Gould & Lewontin (this symposium) argue for pluralism in the explanation of evolutionary change. They believe that the 'adaptationist' programme is methodologically weak, and that recent developments have cast doubt upon the general applicability of the Darwinian theory. It is my purpose here, in discussing contemporary genetic variation, not only to reject this argument, but also to attempt a Darwinian synthesis of recent observations and experiments.

Although some of the modern challenges to the theory of natural selection have concerned the long-term evolution of molecules, the heat of battle has been greatest in discussions of the diversity within populations. Advocates of natural selection have often argued, or assumed, that the most important factor in maintaining this diversity is heterozygous advantage (see, for example, Ford 1971). Thus it has come about that some who have found evidence against the general occurrence of heterozygous advantage have taken it as evidence against the general involvement of natural selection. However, this is not a valid progression.

The great stength of the Darwinian theory is its emphasis on the mechanisms of adaptation and selection, mechanisms that can be studied by observation and experiment. When we examine the agents that are likely to have the most widespread selective effects, we find that nearly every one works in a way liable to produce frequency-dependent selection of the kind that favours rare phenotypes and hinders common ones. Thus, even if the extent, pattern, and behaviour of genetic variation in natural populations argue against the view that the majority of variants are maintained by heterozygous advantage, we are not obliged to abandon the theory of natural selection, even to the extent demanded by the advocates of pluralism. In attempting to make this point, I shall first review very briefly the facts of intraspecific variation, then consider the arguments against heterozygous advantage, and against selective neutrality. Finally, I shall call attention to the evidence that frequency-dependent selection is a powerful, perhaps a dominant, factor in maintaining genetic diversity.

2. Genetic variation within species

It has long been known that most populations of animals and plants contain large amounts of genetic variation. The rapid responses of many organisms to artificial selection for metrical characters imply a ready supply of segregating alleles, and the common occurrence of polymorphisms for colour, pattern and morphology is well documented (see, for example, Sheppard 1975; Berry 1977). The discovery that some of the allelic variation in structural genes can be revealed by electrophoresis has generated a large number of surveys estimating the proportion of enzyme loci that are polymorphic (for recent reviews, see Powell 1975; Selander 1976; Nevo 1978). With conventional acrylamide and starch-gel electrophoreses, the average proportion of enzymes that are detectably polymorphic in the higher animals and plants so far surveyed is about 27%, and the average heterozygosity is about 7.5%. Bacteria are also polymorphic (Milkman 1973). The enzymes used in these surveys do not, of course, represent a random sample of all loci. There are clear differences among the enzymes themselves in the frequency of polymorphism (for example, phosphoglucose isomerase is very often polymorphic, malate dehydrogenase more rarely so), and structural proteins seem to be less polymorphic than enzymes. There are also wide differences in the degree of polymorphism between species, and between higher taxonomic categories.

Recently, by varying the pH and concentration of acrylamide in gels, and by the use of tests for heat stability, even more enzyme variation has been revealed (Cochrane 1976; Coyne 1976; Johnson 1977). Studies on the xanthine dehydrogenase of *Drosophila pseudoobscura* have produced a dramatic increase in the number of 'alleles' from 6 to 37 (Singh *et al.* 1976). However, these additional variants have not yet been positively assigned to the structural locus for the enzyme. Sampsell (1977) found three heat-stability variants of alcohol dehydrogenase, which were mapped at the structural locus, but their combined frequency in natural American populations of *D. melanogaster* was less than 2 %. She found no such variants of α-glycerophosphate dehydrogenase.

Whatever may be the final effect of these new techniques on the estimates of polymorphism, it is clear that conventional electrophoresis fails to reveal a significant proportion of amino acid substitutions (Boyer *et al.* 1972; King & Wilson 1975). Some substitutions, such as isoleucine for valine, are unlikely to have any great effect on either the charge or the general conformation of a protein. It seems probable that for enzymes the proportion of loci that are polymorphic for amino acid substitutions will turn out to be greater than 40 %, the average heterozygosity more than 15 %, and the average number of alleles (with frequencies above 1 %) per polymorphic locus greater than four. The proportion of polymorphic loci, the heterozygosity, and the numbers of alleles are likely to be even greater at the level of the DNA.

It is not yet clear whether genetic variation in metrical characters represents segregation at structural loci. Wilson (1976) has suggested that it is largely due to 'regulatory' alleles that alter the pattern, rather than the quality, of gene expression. The number of abdominal bristles in *D. melanogaster* seems to be partly determined by the number of 'bobbed' (ribosomal RNA) genes, which can rapidly increase or decrease as a result of unequal crossing over (Frankham *et al.* 1978). Alterations in the quantity or disposition of heterochromatin may also affect metrical characters. There are certainly many nucleotide substitutions in the DNA that are not reflected in the amino acids of proteins, a situation exemplified by recent studies on the nucleotide sequences of structural genes (Grunstein & Grunstein 1978; Sim *et al.* 1978; Salser 1978). It is not valid to conclude that these substitutions have no phenotypic effect (Clarke 1970*a*), but such effects are likely to show themselves as quantitative or metrical variation rather than discrete polymorphism.

3. Heterozygous advantage

The evolutionary geneticist is faced with explaining a formidable amount of intraspecific genetic variation. The conventional response to this challenge has been to assume a correspondingly large amount of heterozygous advantage (Ford 1971; Berger 1976). The belief in the importance of heterozygous advantage stems from the frequent occurrence of heterosis, in which hybrids, whether of whole chromosome sets, individual chromosomes, or linked blocks of genes, are larger, more vigorous, and less variable than their parents (Gowen 1952). It has seemed reasonable to extend the phenomenon of heterosis to individual genes, and this extension has been supported by one clear example of single-gene heterozygous advantage (the sickle-cell trait in man; see Allison 1964) and one strong case for it (Warfarin resistance in rats, reviewed by Berry (1977)). It has also been supported by three general arguments:

(i) *The argument that heterozygous advantage evolves*

Barring random genetic drift, if a mutant is to invade a population its heterozygote(s) must have an advantage over the prevailing homozygote(s). During its spread, heterozygotes can evolve an advantage over the mutant homozygote. The selection of dominance modifiers should act so that each allele becomes dominant for its favourable characteristics and recessive for its unfavourable ones. Thus the heterozygotes may come to enjoy the best of both worlds (Sheppard 1975).

(ii) *The argument that hybrid molecules make better enzymes*

Fincham (1972) has drawn attention to a possible resemblance between heterozygous advantage and intracistronic complementation, in which two or more defective mutant polypeptides produce a functioning enzyme when they combine to form a dimer or polymer. It seems that aggregation can correct the inadequacies of one or both polypeptides. If this kind of correction is not limited to severely defective mutants, a dimer or polymer between two naturally occurring allelic variants might have properties superior to those of its parental types. Fincham argues that this kind of complementation may be a general cause of enzyme polymorphism.

(iii) *The argument that heterozygotes are 'better buffered'*

This argument follows Haldane (1954) and Lerner (1954), who suggested that heterozygosity is in itself advantageous because it provides a buffer against genetic or environmental disturbances of the phenotype. A heterozygote may gain from having two (or three) enzymes instead of one. Although heterozygotes might not be superior in a single unchanging environment, they could be at an advantage when the environment varies in space or time. This argument has been extended by Gillespie & Langley (1974; Gillespie 1977), who have derived conditions for polymorphism at a locus in terms of the degree of buffering, expressed as the variance of fitness.

Although the arguments listed above are persuasive, there are equally powerful counter-arguments, and a good deal of evidence, suggesting that heterozygous advantage does not provide a satisfactory general explanation for widespread polymorphism.

In the first place the apparent correspondence between heterozygous advantage and heterosis may be misleading. Heterosis between blocks of genes can be satisfactorily explained in terms of dominance at individual loci rather than overdominance (Gowen 1952).

Secondly, the argument that heterozygous advantage evolves, although probably often correct, carries with it the consequence that once this state had arisen it should subsequently devolve. Initially during the spread of a mutant gene, it will occur predominantly in the heterozygous condition, and the heterozygotes will be more prone than mutant homozygotes to selection for improvement. However once the polymorphism has been established the mutant homozygotes will be common, and will be subject in their turn to modification. This will tend to destabilize the polymorphism. The stability of a polymorphism can be measured by the slope of the curve of Δq against q at equilibrium (i.e. the value of $\mathrm{d}\Delta q/\mathrm{d}q$ when $\Delta q = 0$). This has been called the 'resilience' by Cavalli-Sforza & Bodmer (1971). For a population in Hardy–Weinberg equilibrium with three genotypes AA, Aa and aa, whose frequencies are p^2, $2pq$ and q^2 and whose selective values are $1-s$, 1 and $1-t$ respectively, the resilience is $-st/(s+t-st)$. Its magnitude decreases if s or t are decreased. The greater the value of q, the stronger will be the selection to reduce the value of t. We expect selection to reduce progressively the magnitude of the resilience until the mutant homozygote can be improved no further. If the polymorphism persists despite the reduction in resilience, another phenomenon may become important. This is gene duplication, which should be favoured because it allows the heterozygosity to be fixed, avoiding the production of relatively disadvantageous homozygotes (Spofford 1969). Spontaneous duplication seems, in evolutionary terms, to be a frequent occurrence (perhaps as frequent as gene mutation) and any polymorphism involving strong heterozygous advantage is therefore likely to be a transient condition. Presumably duplication carries a metabolic cost, so that polymorphisms maintained by weak heterozygous advantage might be expected to persist longer, were they not the ones most likely to be lost by random genetic drift or environmental disturbances.

The argument that hybrid proteins make better enzymes cannot be accepted as a general explanation of polymorphism because monomeric enzymes are, on average, more often polymorphic than those composed of dimers or higher polymers (Harris *et al.* 1977; Ward 1977).

The argument that heterozygotes are 'better buffered' is unsatisfactory because it depends upon the application of observations about heterosis (in particular the lower variance of metrical characters and/or fitness in hybrids) to arguments about single loci. As we have already noted, this extrapolation is not necessarily valid, particularly when there appears to be no firm basis in observation or experiment for the postulated homeostatic advantage of heterozygosity at the level of a single locus.

Doubts about the efficacy of heterozygous advantage have been greatly intensified by three recent observations. The first is that genetic variability does not seem to be dramatically reduced in self-fertilizing species (Selander 1976) or in haploid bacteria (Milkman 1975), neither of which contain many heterozygotes. The second is the observation by Allendorf (1978) that following tetraploidy in two independent groups of fish, the loss of duplicate loci is significantly more frequent among the enzymes that show the higher mean levels of heterozygosity (averaged over all the animal taxa so far studied; Powell 1975). Allendorf treats this finding as evidence in favour of the 'neutral' theory, but strictly it argues only against heterozygous advantage at the loci involved. The third observation comes from a theoretical study by Lewontin *et al.* (1978), who point out that heterozygous advantage can maintain more than three alleles at a locus only under very restricted conditions, which become rapidly more severe as the number of alleles increases (because the number of genotypes grows disproportionately with the number of alleles, n alleles producing $n(n+1)/2$ genotypes). An average of five alleles (15 genotypes) per polymorphic locus would present an almost insuperable difficulty to proponents of heterozygous advantage. Even if the extremely stringent conditions for balance were fulfilled, they would lead to a uniform distribution of gene frequencies, with all alleles at a locus in approximately equal proportions. This is not found in nature.

4. Neutrality

The 'neutral' theory of protein evolution has recently been well summarized by Kimura (1977). It was originally formulated (Kimura 1968; King & Jukes 1969) at a time when there seemed to be two serious difficulties for the theory of natural selection. The first was the apparent uniformity of evolutionary rates among diverse lineages of the same protein. The second was the supposed problem of 'genetic load' in the simultaneous selection of many amino acid substitutions during evolution, and in the maintenance of polymorphism. It has since transpired that both difficulties are illusory, at least in their strength. Evolutionary rates are not constant (Langley & Fitch 1974) and although it has been argued that they are more nearly constant than would be expected under selection, the argument has little power because the theory of selection makes no firm prediction about the degree of constancy. Genetic load is no longer a problem (Sved *et al.* 1967; Wallace, 1970, 1975; Clarke 1972, 1973). The concept in its strong form treated relative selective values as if they were absolute fitnesses, a procedure that is certainly not generally valid. Thus the 'neutral' theory is, in a real sense, redundant. Nevertheless it has been developed to a point where it can satisfactorily explain many of the observations on the evolution of proteins, and on the pattern and extent of contemporary enzyme polymorphisms. Of course, the selective theory can do the same, often with more economy (Clarke 1970*a*, *b*; Richmond 1970).

There have been various attempts to discriminate between the 'neutral' and selective theories on the basis of patterns in gene and genotype frequencies, numbers of alleles, measures of heterozygosity, and so on, both within and between species. They have not been successful (Ewens 1977; Ayala 1977; Yamazaki 1977). It seems that almost any pattern can be accommodated by either theory if suitable assumptions are made about mutation rates, migration rates, population sizes, selective values and historical factors. However, it is still possible to discriminate experimentally between the two alternatives. There is an asymmetry about this discrimination because the neutral theory is a null hypothesis, implying an absence of systematic effects. The theory of natural selection, in its broadest sense, implies merely a departure from neutrality. Stated thus it cannot be disproved, and has therefore been considered philosophically unsatisfactory. If there were no more to it than this, the criticism might be valid, but in fact the theory of natural selection comprises a number of separate and particular hypotheses about the mechanisms of selection. These hypotheses are potentially subject to refutation by experiment. Thus it is not sufficient merely to search for apparent departures from neutrality, which might have trivial causes. We must seek, and study, actual or potential mechanisms of selection. In enzyme polymorphisms, these mechanisms presumably act through biochemical differences between allozymes; and in the majority of cases so far investigated, such differences have been found (for reviews, see Harris 1975; Clarke 1975*a*; see also below). There are now several experiments showing that the biochemical differences are reflected in differential survival. For example, in *D. melanogaster* the two common variants of alcohol dehydrogenase, fast (*F*) and slow (*S*), apparently differ by only a single amino acid (threonine in *F*, lysine in *S*; D. Thatcher, personal communication). This substitution alters V_{max}, K_m, temperature stability, substrate specificity, and behaviour with inhibitors (Vigue & Johnson 1973; Day *et al.* 1974; D. Thatcher, unpublished). The results of selection experiments with various alcohols rather exactly match expectations based on the substrate specificities measured *in vitro* (Morgan 1975; van Delden *et al.* 1978). Other experiments involving appropriate environmental stresses have given evidence of selective differences between allozymes of amylase (Scharloo *et al.* 1977; Hickey 1977; Hoorn 1978) and esterase (Ghandhi 1978) in *D. melanogaster*. These studies strongly suggest that the selection acts directly on the enzyme locus, rather than on other loci in linkage disequilibrium with it. Bijlsma & van Delden (1977) and Bijlsma (1978), using similar techniques, report selection acting on allozymes of glucose-6-phosphate dehydrogenase and 6-phosphogluconate dehydrogenase. The careful study by Yamazaki (1971) on esterase in *D. pseudoobscura* found no evidence of selection, but in this case the flies were not specifically stressed. In general it seems that whenever enzyme, or other, loci have been exposed to stresses clearly related to their particular functions, selection has been detected. However, among them there has not yet been any clear example of heterozygous advantage (Jones & Yamazaki 1974; Willis *et al.* 1975).

5. Ecological considerations

The arguments against both heterozygous advantage and neutrality are strong enough to encourage a serious search for alternative explanations of widespread polymorphism. Fortunately, we do not have far to seek. It has long been known that frequency-dependent selection, favouring rare genotypes, is potentially capable of maintaining polymorphism (Fisher 1930; Haldane 1932). Only recently, however, has evidence accumulated indicating that this kind of selection is a major force in promoting and supporting genetic diversity (Clarke 1975*b*).

There is an interesting contrast between the sources of evidence for heterozygous advantage and neutrality on the one hand, and for frequency-dependent selection on the other. Support for the first two mechanisms has come mainly either from theoretical studies or from experiments on organisms removed from their natural environment. Support for frequency-dependent selection has come not only from these sources, but also from studies on actual or potential selective agents. The wealth of this evidence does not suggest that there is any inherent weakness in a Darwinian emphasis on the mechanisms of adaptation.

At this point it is worthwhile to remark briefly upon the parallel between genetic polymorphism, which is the stable coexistence of genotypes, and the persistence of communities, which is the stable coexistence of species. The equilibrium of species numbers in a community is generally considered to be the result of density-dependent mortality and fertility. If two genotypes were ever sufficiently distinct for their numbers to be regulated entirely independently, they could be treated ecologically as two non-competing species. The balancing density-dependent factors would then act as frequency-dependent selective agents. If, as seems more likely, the genotypes were not independently regulated, they could be treated ecologically as competing species, which can be maintained in a state of balance by shared density-dependent factors. These factors would again behave as frequency-dependent selective agents (Williamson 1957; Clarke 1972).

The major density-dependent factors affecting population size are likely to be predation, parasitism and competition. It is among these factors that we might expect to find frequency-dependent selective agents.

6. Predation

The remarkable examples of crypsis found among prey species testify to the effectiveness of selection by predators. These protective resemblances not only involve colour and pattern, but also morphology and behaviour (Cott 1940; Robinson 1969). The phenomena of mimicry are equally striking (Wickler 1968; Vane-Wright 1976; Turner 1977). Some palatable species of butterflies, such as the swallowtails *Papilio dardanus* and *P. memnon*, are polymorphic, with different morphs mimicking different distasteful models (Clarke & Sheppard 1971). In the polymorphic mimic *Pseudacrea eurytus*, the frequencies of the mimetic morphs are nicely adjusted to the numbers of their respective models (Sheppard 1959). This adjustment implies that the selection on the polymorphism is frequency-dependent. When a mimetic form is relatively rare compared with its model, predators encountering its particular visual pattern meet distasteful models more often than they meet palatable mimics. Their avoidance of the pattern will therefore be strengthened. In the converse sense, predators more often meet palatable mimics, and unless the models are very distasteful the avoidance will be weakened. These effects have been demonstrated experimentally (Brower 1960; O'Donald & Pilecki 1970; Ford 1971; Pilecki & O'Donald 1971).

Even in the absence of mimicry, predators select in a frequency-dependent manner (Clarke 1962, 1969). They concentrate disproportionately on common varieties of prey, and overlook rare ones. With respect to colour variants of the prey, this tendency has been shown experimentally in rudd and minnows predating corixid bugs (data of Popham 1941; 1942; reanalysed by Clarke 1962), sticklebacks predating freshwater isopods (Maskell *et al.* 1977), wild passerine birds predating artificial baits (Allen & Clarke 1968; Allen 1972*a*, *b*, 1974, 1975; Horsley 1978), captive quails predating artificial baits (Manly *et al.* 1972; Cook & Miller 1977), and chicks predating coloured crumbs (Fullick & Greenwood 1979). In Allen's (1974, 1975) experiments 35 out of 36 independent groups of wild birds showed a significant tendency to overpredate the common or familiar variety. Soane & Clarke (1973) found the same tendency in mice hunting by smell for baits scented with peppermint and vanilla. There is also evidence of frequency-dependent predatory behaviour, or its equivalents, in various insects (Lawton *et al.* 1974; Hassell *et al.* 1977), dog-whelks (Murdoch 1969), trout (Bryan 1973), guppies (Murdoch *et al.* 1975), gulls (Rabinowitch 1968), pigeons (Murton 1971), crows (Croze 1970) and jays (Coppinger 1969).

Holling (1965, 1966) has investigated the functional responses of predators to changes in prey density, and Horsley (1978) has shown that Holling's theory elegantly explains the observed behaviour of predators to their polymorphic prey.

The widespread occurrence of frequency-dependent selection by predators (which has been called apostatic selection; Clarke 1962) is now established beyond doubt. Whenever it occurs it will contribute to the maintenance of polymorphisms for colour, pattern or smell among the prey. It may also act to promote diversity of morphology (Clarke & Sheppard 1962) and behaviour (Humphries & Driver 1967). The possibility of selection for polymorphism of sounds remains to be explored.

In the circumstances we may ask why, if apostatic selection is so common, are there not *more* polymorphisms for colour, pattern, smell, morphology and behaviour. There are several answers to this question. The first is that such polymorphisms may be more frequent than we realize. Some part of the variance in most metrical characters is due to the segregation of genes (conventional or otherwise) and Allen (1972*a*) has shown, using a quasi-continuous gradation of colour in artificial prey, that predation can act on these characters to increase their variance (in other words that the selection can be disruptive *sensu* Thoday (1972) as well as frequency-dependent). Allen made a population of baits containing intergrades between green and brown, in nine steps. He arranged that the intermediate colour was the commonest, and that the proportions of the colours declined with their difference from the intermediate, approximating a normal distribution. Wild birds overpredated the modal (average) colour and underpredated the extremes. Thus organisms that show continuous variation in colour, pattern, or morphology and are polymorphic at the level of the genes may owe some of their variability to frequency-dependent predation (see also Sheppard 1975).

There are other ways in which we may underestimate the degree of polymorphism. Diversity in smells may go undetected because of our insensitive noses. The story that tracker dogs are confused only by identical twins suggests that the human species might be genetically variable in this respect. Some kinds of polymorphism, despite being obvious, may be neglected in discussions of predation because of preconceived notions about their *raison d'être*. An example of this category is the whole range of sexual dimorphism. It seems entirely reasonable that selection by predators should act to accentuate sexual differences in colour, pattern, smell, morphology or behaviour.

Even making allowance for these arguments, there are still many prey species that are apparently uniform, and it is necessary to remember that in some circumstances, predators can select *against* diversity. Obvious examples are distasteful and warningly coloured species. It seems that selection for uniformity can occur even in the absence of distastefulness. Allen (1972*a*, *b*) and Horsley (1978) have found that when palatable baits are exposed at very high density to wild birds, the predators tend to take disproportionately many of the *rare* colours. Mueller (1971) reported a similar pattern of predation by hawks attacking flocks of pigeons. One possible explanation of this phenomenon is that, when several prey are simultaneously within the visual field of the predator, rare colours 'stand out' against the background of the common ones. I am grateful to D. T. Parkin for pointing out that species of birds habitually flying in flocks tend more often to be uniform in colour. The same may be true of fishes that habitually swim in schools The likelihood of selection for polymorphism depends upon the density of the prey.

Of course there are many other possible constraints upon polymorphism. For example, the requirements of crypsis or of mate recognition, the mechanics of locomotion, or the demands of metabolism may set limits to the degree of phenotypic variability.

An interesting extension of frequency dependence in predator–prey interactions has been suggested by Paulson (1973). He has argued that prey species, particularly those quick to learn, can exert frequency-dependent selection on their predators. They may come to associate a particular visual (or olfactory or behavioural) pattern with the predator, and learn to avoid it. This would confer an advantage on any predator that departed from the visual (or olfactory or behavioural) norm. Paulson found support for this hypothesis in his studies of birds of prey. Avian predators hunting prey that are relatively intelligent and visually acute seem to be more often polymorphic for colour than those hunting relatively unintelligent and dim-sighted prey.

7. Parasitism

The great array of complex arrangements that organisms have evolved as defences against parasites, from the possession of hairy leaves to the development of antibodies, shows the importance of parasitism in affecting the course of evolution. In man a single group of related diseases, the malarias, is responsible for selection acting on at least three polymorphisms (haemoglobin, glucose-6-phosphate dehydrogenase and the Duffy blood groups (Motulksy 1975; Miller *et al.* 1975)), and probably on several more, including the genes regulating levels of ATP (Motulsky 1975) and perhaps some of the HLA alleles (Piazza *et al.* 1973). If each major class of disease acts on a similar crop of variants, disease-related polymorphism may be a very large component of human biochemical diversity. The same logic applies to other organisms.

It was Haldane (1949) who first suggested that parasites could maintain polymorphisms in their hosts. Since then we have learned a good deal more about the nature of host–parasite interactions, and the case has become very much stronger.

If a host is initially uniform in its genetic composition, and if a parasite is finely adapted to the conditions of life in the host, then even a small biochemical change in the host might confer some degree of resistance to the parasite. Consequently as Haldane (1949) pointed out, a host with a rare biochemical phenotype can gain a selective advantage. It follows, if the phenotype is genetically determined, that the corresponding genotype will increase in frequency. As its frequency grows the advantage will decline, because natural selection will favour variant parasites that can more effectively attack the new type of host. The appropriate phenotypes in the parasite will then begin to spread. Their subsequent behaviour will depend upon their relative success in attacking the original type of host. If they are able equally to infect both host genotypes, and are as fit as the original phenotypes in other respects, they will spread to fixation. If, however, an increased success in attacking the new host genotype carries with it a reduced ability to attack the old, a 'feedback-loop' will be established, in which polymorphisms can be actively maintained in both host and parasite (Clarke 1976).

This sequence of events will take place if the following conditions are fulfilled:

(i) that there is genetic variation in the ability of hosts to resist parasites;

(ii) that there is variation in the ability of parasites to attack hosts;

(iii) that there are specific relations between these two types of variation (particularly that an increased ability on the part of the host to resist one type of parasite sometimes reduces its ability to resist another);

(iv) that parasites kill, or otherwise debilitate, hosts.

The evidence for conditions (i), (ii) and (iv) is very extensive (Day 1974; Clarke 1976) and need not be repeated here. The evidence for condition (iii) is, however, particularly interesting. Precise relations between the genotypes of host and parasite were first described by Flor (1956) in his studies on flax and flax rust. He found that for each gene conditioning the reaction of the host, there is a specific gene conditioning pathogenicity in the parasite. Such gene-for-gene relations have now been found in seven plant–fungus combinations (Day 1974), and also between wheat and an insect parasite, *Mayetiola destructor* (Hatchett & Gallun 1970). Mode (1958) and Person (1966) have pointed out that this kind of system can maintain polymorphisms in both interacting species (for a practical application of these arguments, see Barrett 1978).

It has been suggested that gene-for-gene relations may also occur between vertebrate hosts and their parasites (Damian 1964, 1979). A parasite in an immunologically competent host could protect itself from attack by reducing its antigenic disparity with the host, and 'mimicking' the host's antigens. This would produce selection in favour of antigenically variant hosts that could damage the parasite. The literature on antigen sharing between host and parasite is reviewed by Damian (1979). Although it is often uncertain whether parasites have gained their cross-reacting antigens by the natural selection of genetic variants, rather than directly from their hosts (as certainly happens with those viruses that wrap themselves in host cell membranes), the presence of these antigens, regardless of their origin, could still provoke polymorphisms in the hosts. The existence of antigenic polymorphisms associated with differential resistance to disease (Vogel 1975; Bodmer & Bodmer 1978) adds strength to these arguments. All the current theories about the functions and evolution of the major histocompatibility loci in vertebrates imply frequency-dependent selection by parasites (Bodmer & Bodmer 1978). It is notable that some of the loci in the human HLA complex have more than 20 alleles, with skewed distributions of frequencies. There seems to be no difficulty, either practical or theoretical, in the maintenance of large numbers of alleles by frequency-dependent selection.

Although specific relationships between the genes of vertebrates and their parasites have been more thoroughly discussed at the immunological level, it seems inevitable that they will occur at other levels as well. The work of Fenner & Ratcliffe (1965) has shown the evolution of a delicate mutual adjustment, that is certainly not entirely immunological, between the rabbit and the myxoma virus.

There is a second way by which parasites can act to maintain polymorphisms in their hosts. Even when a parasite is unable to overcome the resistance conferred by a new host genotype (i.e. if the genotype confers a 'general' rather than a specific resistance to the parasite) a polymorphism can still be maintained if the numbers of parasites are to some extent determined by the availability of susceptible hosts. When the gene for general resistance is rare there will be many parasites, and relatively strong selection in favour of resistant genotypes. When the gene is common there will be fewer parasites, and weaker selection in favour of resistant genotypes. The selection on the host will therefore be frequency-dependent. If the resistance gene has pleiotropic effects that are disadvantageous, there will be some frequency at which the advantageous and disadvantageous effects cancel each other. This will be a stable point of equilibrium.

Mathematical models of genetic interactions between hosts and parasites have been made by Jayakar (1970), Cook (1971), Yu (1972) and Clarke (1976). Jayakar simulated a model of 'general virulence', in which a new virulent genotype was able to attack previously resistant genotypes of host, as well as previously susceptible ones. The numbers of parasites and hosts were held constant and, as might be expected, there was no balanced polymorphism. Cook and Yu simulated a situation in which there were specific genetic relations between parasite and host (one host allele giving resistance to the effects of one allele in the parasite, but susceptibility to the effects of the other, the alternative host allele giving resistance and susceptibility in the opposite sense). Cook's haploid model did not produce a balanced polymorphism, and Yu's diploid model only did so when there was heterozygous advantage for virulence in the parasite. These models, however, did not include the capacity of parasites to evolve virulence within the host. When this factor is introduced, polymorphism becomes almost inevitable, and it is maintained even when both host and parasite are haploid (Clarke 1976). There is either a stable equilibrium or a stable limit cycle.

A model of 'general' resistance also suggests that the conditions for polymorphism are robust (Clarke 1976). The breadth of these conditions in both models suggests that frequency-dependent selection by parasites may be a potent factor in maintaining biochemical diversity.

8. Competition

When members of the same species are competing for some limited resource, they must necessarily interact. It is convenient for our purposes to divide the possible interactions into two classes: direct interactions between individuals (mating, aggression and so on), and indirect interactions, mediated through the resource. The classes are not, of course, entirely discrete.

(1) *Direct interactions*

The self-incompatibility polymorphisms found in many plants are classic examples of frequency-dependent selection (reviewed by Lewis (1949) and Wright (1969)). In a population containing only alleles that confer self-sterility, individuals with rare alleles should enjoy a relative advantage because their pollen ought more often to produce successful fertilizations. An extreme case of self-incompatibility is the existence of sex itself. The frequency-dependent factors involved in the maintenance of the sex ratio are discussed by Williams (this symposium).

Similar in principle, but less extreme in effect, is negative assortative mating, a disproportionate tendency for unlike individuals to mate together. It has been reported to occur in the scarlet tiger moth (Sheppard 1952; Sheppard & Cook 1962) and in the arctic skua (O'Donald *et al.* 1974; Davis & O'Donald 1976). Mating preferences tending to favour rare phenotypes have been reported in

Drosophila (Petit 1951; Ehrman 1966; Spiess 1968), flour beetles (Sinnock 1970), butterflies (Levin 1973), parasitic wasps (Grant *et al.* 1974) and milkweed beetles (Eanes *et al.* 1977). The mechanisms underlying these preferences are not yet clear, so that it is difficult to say with certainty what cues, characters, and genes are involved in the frequency-dependent selection. There is, however, evidence for both olfactory and auditory cues (Ayala & Campbell 1974; Ehrman & Parsons 1976; Ewing 1978).

Other behavioural characters, including many of the 'strategies' discussed later in this symposium, are probably subject to frequency-dependent advantages and disadvantages. For example, a tendency to aggression may be individually advantageous if it is rare, but disadvantageous if it is common.

(ii) *Indirect interactions*

If two genotypes differ in such a way that one genotype is better at exploiting one component of the environment and the other is better at exploiting another component, the genotypes can be regarded as occupying different 'ecological niches'. Levene (1953) first investigated theoretically the effects of differential selection in two or more niches, and showed that polymorphism is maintained if the weighted harmonic mean selective value of the heterozygotes in the various niches is higher than those of the homozygotes. It is necessary to emphasize that although Levene's model, and those derived from it (reviewed by Hedrick *et al.* 1976; Felsenstein 1976), state the conditions for polymorphism in terms of constant selective values within the niches, as far as the whole population is concerned the selection is frequency-dependent. The models assume that the numbers of organisms in each niche are independently regulated after selection has taken place. It is this, necessarily density-dependent, regulation that introduces the frequency-dependent component of selection. When it is absent, there is no polymorphism without conventional heterozygous advantage (Dempster 1955). Selection in multiple niches is not an alternative to frequency-dependent selection, as implied by some authors (for example, Lewontin *et al.* 1978), but a way of generating it.

Levene's model has somewhat stringent conditions for polymorphism, but these conditions become relaxed if the genotypes tend to choose their appropriate niches (Maynard Smith 1966) or if there is 'facilitation', where the metabolic products of one genotype favour the other. The conditions also seem to be more generous if the density-dependent regulatory factors are the actual selective agents (Clarke 1972, and unpublished).

In this context Clarke & Allendorf (1979) have pointed out that polymorphisms in enzymes can be subject to frequency-dependent selection exerted by the availability of their substrates. This conclusion follows directly from the theory of enzyme kinetics. Alternative allozymes often differ in their kinetic properties (Harris 1975; Clarke 1975*a*), and it seems that in all the cases so far examined when one allozyme has a higher V_{max} than another, it also has a higher K_m (Clarke & Allendorf 1979; see also Pollock 1965). This means that the relative velocity of the reaction catalysed by one allozyme compared with that catalysed by the other will vary with the concentration of the substrate. If fitness is a monotonically increasing function of reaction velocity, if the rate of the reaction determines the uptake of substrate (or a precursor of it) from the environment, and if the initial concentration of substrate is sufficiently high, then the presence of larger numbers of individuals with the 'high V_{max}' enzyme will more rapidly deplete the supply of substrate, and so will relatively favour individuals with the 'low V_{max}' enzyme. Thus the selection will be frequency-dependent. Computer simulations show that, given appropriate concentrations of substrate, and numbers of organisms, the effect can be strong. It is possible that this mechanism is a common cause of polymorphism among enzymes with external substrates. Whether the system can easily maintain more than two alleles, however, remains to be discovered.

9. Frequency-dependent selection whose mechanism is unknown

There are now many reports of experiments in which the selective values of genotypes and karyotypes have been found to be negatively correlated with their frequencies. These experiments have been reviewed by Petit (1968, 1973), Murray (1972), Ayala & Campbell (1974), Anxolabéhère (1976) and de Benedictis (1978*a*, *b*). The mechanisms producing the frequency-dependent selection are unknown. Experiments involving larval competition in *Drosophila* as, for example, in the careful studies by Bundgaard & Christiansen (1972), evidently cannot involve selection by predators, and the frequency dependence has usually been attributed to the effects of multiple niches, or of 'facilitation' between genotypes. The possible role of parasites has, I believe, never been considered.

Frequency-dependent selection acting on enzyme genotypes has been reported by Kojima & Yarbrough (1967), Kojima & Tobari (1969), and, under more strictly controlled conditions, by Morgan (1976) and Ghandhi (1978). The negative results obtained by Yamazaki (1971) and Dolan & Robertson (1975) may perhaps have been due to conditions that were insufficiently competitive. In the experiments that have given positive results, it is not certain that the selection is acting directly on the enzyme loci rather than on other loci closely linked to them. However, it is clear that frequency-dependent selection acts on *some* loci. In this respect, there is an important difference between the evidence for heterozygous advantage and the evidence for frequency dependence. In the former case, linkage disequilibrium can generate heterosis and produce spurious evidence of heterozygous advantage. In the latter, linkage disequilibrium can only mislead us about the target of the selection, not about its nature.

10. Conclusions

It seems that a general explanation of widspread polymorphism in terms of frequency-dependent selection suffers none of the difficulties raised by an interpretation in terms of heterozygous advantage. Frequency-dependent polymorphisms are not expected to devolve, nor to be lost by gene duplication. They are compatible with our knowledge about the structure and function of proteins, and indeed seem logically to follow from the kinetics of enzymes. They are expected to occur in haploid and self-fertilizing organisms. They do not conflict with Allendorf's (1978) data on gene loss. They can support many alleles at a locus, even with skewed distributions of allelic frequencies. Many of the most important agents of selection are known or expected to act in a frequency-dependent manner, and the effects of this action are likely to be manifest in metrical variation as well as in phenotypic polymorphism. Consequently we expect frequency-dependent selection to maintain variability in the non-coding regions of the DNA as well as in structural genes.

The accumulation of evidence for the theory of frequency dependence, and its general consistency with the facts of variation, give no encouragement to those who argue that we should abandon the Darwinian canon, even to the extent of a 'pluralistic' compromise.

I am very grateful to Dr F. W. Allendorf and Dr A. G. Clarke for critically reading the manuscript, and to the Science Research Council and the Medical Research Council for financial support.

I had hoped to contribute a predecessor of this paper to the book commemorating the late Professor Ove Frydenberg, but I was prevented by illness from doing so. I should therefore like to dedicate this essay to the memory of a humorous friend, a powerful intelligence, and an outstanding population geneticist.

References (Clarke)

Allen, J. A. 1972*a* Apostatic selection: the responses of wild passerines to artificial polymorphic prey. Ph.D. thesis, University of Edinburgh.

Allen, J. A. 1972*b* Evidence for stabilizing and apostatic selection by wild blackbirds. *Nature, Lond.* **237**, 348–349.

Allen, J. A. 1974 Further evidence for apostatic selection by wild passerine birds: training experiments. *Heredity, Lond.* **33**, 361–372.

Allen, J. A. 1975 Further evidence for apostatic selection by wild passerine birds: 9:1 experiments. *Heredity, Lond.* **36**, 173–180.

Allen, J. A. & Clarke, B. 1968 Evidence for apostatic selection by wild passerines. *Nature, Lond.* **220**, 501–502.

Allendorf, F. W. A. 1978 Protein polymorphism and the rate of loss of duplicate gene expression. *Nature, Lond.* **272**, 76–78.

Allison, A. C. 1964 Polymorphism and natural selection in human populations. *Cold Spring Harb. Symp. quant. Biol.* **29**, 139–149.

Anxolabéhère, D. 1976 Heterosis, overdominance and frequency-dependent selection in *Drosophila melanogaster* at the *sepia* locus. *Evolution* **30**, 523–534.

Ayala, F. J. 1977 Protein evolution in different species: is it a random process? In *Molecular evolution and polymorphism* (ed M. Kimura), pp. 73–102. Mishima, Japan: National Institute of Genetics.

Ayala, F. J. & Campbell, C. A. 1974 Frequency-dependent selection. *A. Rev. Ecol. Syst.* **5**, 115–138.

Barrett, J. A. 1978 A model of epidemic development in variety mixtures. In *Plant disease epidemiology* (ed. P. R. Scott & A. Bainbridge), pp. 129–137. Oxford: Blackwell.

Berger, E. 1976 Heterosis and the maintenance of enzyme polymorphism. *Am. Nat.* **110**, 823–839.

Berry, R. J. 1977 *Inheritance and natural history*. London: Collins.

Bijlsma, R. 1978 Polymorphism at the G6PD and 6PGD loci in *Drosophila melanogaster*. II. Evidence for interaction in fitness. *Genet. Res., Camb.* **31**, 227–237.

Bijlsma, R. & van Delden, W. 1977 Polymorphism at the G6PD and 6PGD loci in *Drosophila melanogaster*. I. Evidence for selection in experimental populations. *Genet. Res.* **30**, 221–236.

Bodmer, W. F. & Bodmer, J. G. 1978 Evolution and function of the HLA system. *Br. Med. Bull.* **34**, 309–316.

Boyer, S. H., Noyes, A. N., Timmons, C. F. & Young, R. A. 1972 Primate haemoglobins and evolutionary patterns. *J. hum. Evol.* **1**, 515–543.

Brower, J. V. Z. 1960 Experimental studies of mimicry. IV. The reactions of starlings to different proportions of models and mimics. *Am. Nat.* **94**, 271–282.

Bryan, J. E. 1973 Feeding history, parental stock and food selection in rainbow trout. *Behaviour* **45**, 123–153.

Bundgaard, J. & Christiansen, F. B. 1972 Dynamics of polymorphisms: I. Selection components in an experimental population of *Drosophila melanogaster*. *Genetics, Princeton* **71**, 439–460.

Cavalli-Sforza, L. L. & Bodmer, W. F. 1971 *The genetics of human populations*. San Francisco: W. H. Freeman.

Clarke, B. 1962 Balanced polymorphism and the divergence of sympatric species. In *Taxonomy and geography* (ed. D. Nichols), pp. 47–70. Oxford: Systematics Association.

Clarke, B. 1969 The evidence for apostatic selection. *Heredity, Lond.* **24**, 347–352.

Clarke, B. 1970*a* Darwinian evolution of proteins. *Science, N.Y.* **168**, 1009–1011.

Clarke, B. 1970*b* Selective constraints on amino-acid substitutions during the evolution of proteins. *Nature, Lond.* **228**, 159–160.

Clarke, B. 1972 Density-dependent selection. *Am. Nat.* **106**, 1–13.

Clarke, B. 1973 Mutation and population size. *Heredity, Lond.* **31**, 367–379.

Clarke, B. 1975*a* The contribution of ecological genetics to evolutionary theory: detecting the direct effects of natural selection on particular polymorphic loci. *Genetics, Princeton* **79**, 101–113.

Clarke, B. 1975*b* Frequency-dependent and density-dependent natural selection. In *The role of natural selection in human evolution* (ed. F. M. Salzano), pp. 187–200. Amsterdam: North-Holland.

Clarke, B. 1976 The ecological genetics of host–parasite relationships. In *Genetic aspects of host–parasite relationships* (ed. A. E. R. Taylor & R. Muller), pp. 87–103. Oxford: Blackwell.

Clarke, B. & Allendorf, F. W. 1979 Frequency-dependent selection due to kinetic differences between allozymes. *Nature, Lond.* (In the press.)

Clarke, C. A. & Sheppard, P. M. 1962 Disruptive selection and its effect on a metrical character in *Papilio dardanus*. *Evolution* **16**, 214–226.

Clarke, C. A. & Sheppard, P. M. 1971 Further studies on the mimetic butterfly *Papilio memnon* L. *Phil. Trans. R. Soc. Lond.* B **263**, 35–70.

Cochrane, B. J. 1976 Heat stability variants of esterase-6 in *Drosophila melanogaster*. *Nature, Lond.* **263**, 131–132.

Cook, L. M. 1971 *Coefficients of natural selection*. London: Hutchinson University Library.

Cook, L. M. & Miller, P. 1977 Density-dependent selection on polymorphic prey – some data. *Am. Nat.* **111**, 594–598.
Coppinger, R. P. 1969 The effect of experience and novelty on avian feeding behaviour with reference to the evolution of warning coloration in butterflies. I. Reaction of wild caught and adult blue jays to novel insects. *Behaviour* **35**, 45–60.
Cott, H. B. 1940 *Adaptive coloration in animals*. London: Methuen.
Coyne, J. A. 1976 Lack of genic similarity between two sibling species of *Drosophila* as revealed by varied techniques. *Genetics, Princeton* **84**, 593–607.
Croze, H. 1970 Searching images in carrion crows. *Z. Tierpsychol.* **5**, 1–85.
Damian, R. T. 1964 Molecular mimicry: antigen sharing by parasite and host and its consequences. *Am. Nat.* **98**, 129–149.
Damian, R. T. 1979 Molecular mimicry in biological adaptation. In *Host–parasite interfaces at population, individual and molecular levels* (ed. B. B. Nickol). New York: Academic Press. (In the press.)
Davis, J. W. F. & O'Donald, P. 1976 Estimation of assortative mating preferences in the Arctic Skua. *Heredity, Lond.* **36**, 235–244.
Day, T. H., Hillier, P. C. & Clarke, B. 1974 The properties of genetically polymorphic isozymes of alcohol dehydrogenase in *Drosophila melanogaster*. *Biochem. Genet.* **11**, 141–153.
De Benedictis, P. A. 1978*a* Are populations characterised by their genes or by their genotypes? *Am. Nat.* **112**, 155–175.
De Benedictis, P. A. 1978*b* Frequency-dependent choice, minority advantages, and frequency-dependent natural selection. *Evolution* **31**, 666–673.
Dempster, E. 1955 Maintenance of genetic heterogeneity. *Cold Spring Harb. Symp. quant. Biol.* **20**, 25–32.
Dolan, R. & Robertson, A. 1975 The effect of conditioning the medium in *Drosophila* in relation to frequency-dependent selection. *Heredity, Lond.* **35**, 311–316.
Eanes, W. F., Gaffney, P. M., Koehn, R. K. & Simon, C. M. 1977 A study of sexual selection in natural populations of the milkweed beetle *Tetraopes tetraophthalmus*. In *Measuring selection in natural populations* (ed. F. B. Christiansen & T. M. Fenchel), pp. 49–64. Berlin: Springer-Verlag.
Ehrman, L. 1966 Mating success and genotype frequency in *Drosophila*. *Anim. Behav.* **14**, 332–339.
Ehrman, L. & Parsons, P. A. 1976 *The genetics of behaviour*. Sunderland, Mass.: Sinauer Associates.
Ewens, W. J. 1977 Selection and neutrality. In *Measuring selection in natural populations* (ed. F. B. Christiansen & T. M. Fenchel), pp. 159–176. Berlin: Spring-Verlag.
Ewing, A. W. 1978 An investigation into selective mechanisms capable of maintaining balanced polymorphisms. Ph.D. thesis, Portsmouth Polytechnic.
Felsenstein, J. 1976 The theoretical population genetics of variable selection and migration. *A. Rev. Genet.* **10**, 253–280.
Fenner, F. & Ratcliffe, F. N. 1965 *Myxomatosis*. London: Cambridge University Press.
Fincham, J. R. S. 1972 Heterozygous advantage as a likely general basis for enzyme polymorphisms. *Heredity, Lond.* **28**, 387–391.
Fisher, R. A. 1930 *The genetical theory of natural selection*. Oxford: Clarendon Press.
Flor, H. H. 1956 The complementary genic systems in flax and flax rust. *Adv. Genet.* **8**, 29–54.
Ford, H. A. 1971 The degree of protection gained by new partial mimics. *Heredity* **27**, 227–236.
Ford, E. B. 1971 *Ecological genetics*, 3rd edn. London: Chapman & Hall.
Frankham, R., Briscoe, D. A. & Nurthen, R. K. 1978 Unequal crossing over at the rRNA locus as a source of quantitative genetic variation. *Nature, Lond.* **272**, 80–81.
Fullick, T. G. & Greenwood, J. J. D. 1979 Frequency-dependent food selection in relation to two models. *Am. Nat.* (In the press.)
Ghandhi, Y. 1978 A study of the esterase-6 enzyme polymorphism in *Drosophila melanogaster*. Ph.D. thesis, University of Nottingham.

Gillespie, J. H. 1977 A general model to account for enzyme variation in natural populations. IV. The quantitative genetics of viability mutants. In *Measuring selection in natural populations* (ed. F. B. Christiansen & T. M. Fenchel), pp. 301–314. Berlin: Springer-Verlag.

Gillespie, J. H. & Langley, C. 1974 A general model to account for enzyme variation in natural populations. *Genetics, Princeton* **76**, 837–848.

Gowen, J. W. (ed.) 1952 *Heterosis*. Ames, Iowa: Iowa State College Press.

Grant, B., Snyder, G. A. & Glessner, S. F. 1974 Frequency-dependent mate selection in *Mormoniella vitripennis*. *Evolution* **28**, 428–438.

Grunstein, M. & Grunstein, J. E. 1978 The histone H4 gene of *Strongylocentrotus purpuratus*. DNA and mRNA sequences at the 5′ end. *Cold Spring Harb. Symp. quant. Biol.* **42**, 1083–1092.

Haldane, J. B. S. 1932 *The causes of evolution*. London: Longmans, Green & Co.

Haldane, J. B. S. 1949 Disease and evolution. *Ricerca scient. Suppl.* **19**, 68–76.

Haldane, J. B. S. 1954 *The biochemistry of genetics*. London: Allen & Unwin.

Harris, H. 1975 *The principles of human biochemical genetics*. Amsterdam: North-Holland.

Harris, H., Hopkinson, D. A. & Edwards, Y. H. 1977 Polymorphism and the subunit structure of enzymes: a contribution to the neutralist–selectionist controversy. *Proc. natn. Acad. Sci. U.S.A.* **74**, 698–701.

Hassell, M. P., Lawton, J. H. & Beddington, J. R. 1977 Sigmoid functional responses by invertebrate predators and parasitoids. *J. Anim. Ecol.* **46**, 249–262.

Hatchett, J. H. & Gallun, R. L. 1970 Genetics of the ability of the hessian fly, *Mayetiola destructor*, to survive on wheats having different genes for resistance. *Ann. ent. Soc. Am.* **63**, 185–194.

Hedrick, P. W., Ginevan, M. E. & Ewing, E. P. 1976 Genetic polymorphism in heterogeneous environments. *A. Rev. Ecol. Syst.* **7**, 1–32.

Hickey, D. A. 1977 Selection for amylase allozymes in *Drosophila melanogaster*. *Evolution* **31**, 800–804.

Holling, C. S. 1965 The functional response of predators to prey density and its role in mimicry and population regulation. *Mem. ent. Soc. Can.* **45**, 1–60.

Holling, C. S. 1966 The functional response of invertebrate predators to prey density. *Mem. ent. Soc. Can.* **48**, 1–86.

Hoorn, A. J. W. 1978 The functional significance of amylase polymorphism in *Drosophila melanogaster*. Ph.D. thesis, University of Utrecht.

Horsley, D. T. 1978 The role of avian predators in maintaining colour polymorphism. Ph.D. thesis, University of Nottingham.

Humphries, D. A. & Driver, P. M. 1967 Erratic display as a device against predators. *Science, N.Y.* **156**, 1707–1708.

Jayakar, S. D. 1970 A mathematical model for interaction of gene frequencies in a parasite and its host. *Theor. Popul. Biol.* **1**, 140–164.

Johnson, G. B. 1977 Hidden heterogeneity among electrophoretic alleles. In *Measuring selection in natural populations* (ed. F. B. Christiansen & T. M. Fenchel), pp. 223–244. Berlin: Springer-Verlag.

Jones, J. S. & Yamazaki, T. 1974 Genetic background and the fitness of allozymes. *Genetics, Princeton* **78**, 1185–1189.

Kimura, M. 1968 Evolutionary rate at the molecular level. *Nature, Lond.* **217**, 624–626.

Kimura, M. 1977 Causes of evolution and polymorphism at the molecular level. In *Molecular evolution and polymorphism* (ed. M. Kimura), pp. 1–28. Mishima, Japan: National Institute of Genetics.

King, J. L. & Jukes, T. H. 1969 Non-Darwinian evolution. *Science, N.Y.* **164**, 788–798.

King, M. C. & Wilson, A. C. 1975 Evolution at two levels: molecular similarities and biological differences between humans and chimpanzees. *Science, N.Y.* **188**, 107–116.

Kojima, K. & Tobari, Y. N. 1969 The pattern of viability changes associated with genotype frequency at the alcohol dehydrogenase locus in a population of *Drosophila melanogaster*. *Genetics, Princeton* **61**, 201–209.

Kojima, K. & Yarbrough, K. M. 1967 Frequency-dependent selection at the esterase-6 locus in *Drosophila melanogaster*. *Proc. natn. Acad. Sci. U.S.A.* **57**, 645–649.

Langley, C. H. & Fitch, W. M. 1974 An examination of the constancy of the rate of molecular evolution. *J. molec. Evol.* **3**, 161–177.

Lawton, J. H., Beddington, J. R. & Bonser, R. 1974 Switching in invertebrate predators. In *Ecological stability* (ed. M. B. Usher & M. H. Williamson), pp. 141–158. London: Chapman & Hall.

Lerner, I. M. 1954 *Genetic homeostasis*. New York: John Wiley.

Levin, M. P. 1973 Preferential mating and the maintenance of the sex-limited dimorphism in *Papilio glaucus*: evidence from laboratory matings. *Evolution* **27**, 257–264.

Levene, H. 1953 Genetic equilibrium when more than one ecological niche is available. *Am. Nat.* **87**, 331–333.

Lewis, D. 1949 Incompatibility in flowering plants. *Biol. Rev.* **24**, 472–496.

Lewontin, R. C., Ginzburg, L. R. & Tuljapurkar, S. D. 1978 Heterosis as an explanation for large amounts of genic polymorphism. *Genetics, Princeton* **88**, 149–170.

Manly, B. F. J., Miller, P. & Cook, L. M. 1972 Analysis of a selective predation experiment. *Am. Nat.* **106**, 719–736.

Maskell, M., Parkin, D. T. & Verspoor, E. 1977 Apostatic selection by sticklebacks upon a dimorphic prey. *Heredity, Lond.* **39**, 83–89.

Maynard Smith, J. 1966 Sympatric speciation. *Am. Nat.* **100**, 637–650.

Milkman, R. 1973 Electrophoretic variation in *E. coli* from natural sources. *Science, N.Y.* **182**, 1024–1026.

Miller, L. H., Mason, S. J., Dvorak, J. A., McGinniss, M. H. & Rothman, I. K. 1975 Erythrocyte receptors for (*Plasmodium knowlesii*) malaria: Duffy blood group determinants. *Science, N.Y.* **189**, 561–563.

Mode, C. J. 1958 A mathematical model for the coevolution of obligate parasites and their hosts. *Evolution* **12**, 138–165.

Morgan, P. 1975 Selection acting directly on an enzyme polymorphism. *Heredity, Lond.* **34**, 124–127.

Morgan, P. 1976 Frequency dependent selection at two enzyme loci in *Drosophila melanogaster*. *Nature, Lond.* **263**, 765–766.

Motulsky, A. G. 1975 Glucose-6-phosphate dehydrogenase and abnormal hemoglobin polymorphisms – evidence regarding malarial selection. In *The role of natural selection in human evolution* (ed. F. M. Salzano), pp. 271–291. Amsterdam: North-Holland.

Mueller, H. C. 1971 Oddity and specific searching image more important than conspicuousness in prey selection. *Nature, Lond.* **233**, 345–346.

Murdoch, W. W. 1969 Switching in general predators: experiments on predator specificity and stability of prey populations. *Ecol. Monogr.* **39**, 335–354.

Murdoch, W. W., Avery, S. & Smyth, M. E. B. 1975 Switching in predatory fish. *Ecology* **56**, 1094–1105.

Murray, J. 1972 *Genetic diversity and natural selection*. Edinburgh: Oliver & Boyd.

Murton, R. K. 1971 The significance of a specific search image in the feeding behaviour of the wood pigeon. *Behaviour* **40**, 10–42.

Nevo, E. 1978 Genetic variation in natural populations: patterns and theory. *Theor. Popul. Biol.* **13**, 121–177.

O'Donald, P. & Pilecki, C. 1970 Polymorphic mimicry and natural selection. *Evolution* **24**, 395–401.

O'Donald, P., Wedd, N. S. & Davis, J. W. F. 1974 Mating preferences and sexual selection in the Arctic Skua. *Heredity, Lond.* **33**, 1–16.

Paulson, D. R. 1973 Predator polymorphism and apostatic selection. *Evolution* **27**, 269–277.

Person, C. 1966 Genetic polymorphism in parasitic systems. *Nature, Lond.* **212**, 266–267.

Petit, C. 1951 Le rôle de l'isolement sexuel dans l'évolution des populations de *Drosophila melanogaster*. *Bull. biol. Fr. Belg.* **85**, 392–418.

Petit, C. 1968 Le rôle des valeurs sélectives variables dans le maintien du polymorphisme. *Bull. Soc. zool. Fr.* **93**, 187–208.

Petit, C. 1973 L'avantage du type rare, facteur de maintien du polymorphisme. *Mém. Soc. zool. Fr.* **37**, 417–441.

Piazza, A., Belvedere, M. C., Bernoco, D., Conighi, C., Contu, L., Curtoni, E. S., Mattiuzi, P. L., Mayr, W., Richiardi, P., Scudeller, G. & Ceppelini, R. 1973 HLA variation in four Sardinian villages under differential selective pressure by malaria. In *Histocompatibility testing 1972* (ed. J. Dausset & J. Colombiani), pp. 73–85. Copenhagen: Munksgaard.

Pilecki, C. & O'Donald, P. 1971 The effects of predation on artificial mimetic polymorphisms with perfect and imperfect mimics at varying frequencies. *Evolution* **25**, 365–370.

Pollock, M. R. 1965 Purification and properties of penicillinases from two strains of *Bacillus licheniformis*: a chemical, physiochemical and physiological comparison. *Biochem. J.* **94**, 666–675.

Popham, E. J. 1941 The variation of the colour of certain species of *Arctocorisa* (Hemiptera, Corixidae) and its significance. *Proc. zool. Soc. Lond.* **111**, 135–172.

Popham, E. J. 1942 Further experimental studies on the selective action of predators. *Proc. zool. Soc. Lond.* **112**, 105–117.

Powell, J. R. 1975 Protein variation in natural populations of animals. *Evolut. Biol.* **8**, 79–119.

Rabinowitch, V. E. 1968 The role of experience in the development of food preferences in gull chicks. *Anim. Behav.* **16**, 425–428.

Richmond, R. C. 1970 Non-Darwinian evolution: a critique. *Nature, Lond.* **225**, 1025–1028.

Robinson, M. H. 1969 Defenses against visually hunting predators. *Evolut. Biol.* **3**, 225–259.

Salser, W. 1978 Globin mRNA sequences: analysis of base pairing and evolutionary implications. *Cold Spring Harb. Symp. quant. Biol.* **42**, 985–1002.

Sampsell, B. 1977 Isolation and genetic characterization of alcohol dehydrogenase thermostability variants occurring in natural populations of *Drosophila melanogaster*. *Biochem. Genet.* **15**, 971–988.

Scharloo, W., van Dijken, F. R., Hoorn, A. J. W., de Jong, G. & Thorig, G. F. W. 1977 Functional aspects of genetic variation. In *Measuring selection in natural populations* (ed. F. B. Christiansen & T. M. Fenchel), pp. 132–147. Berlin: Springer-Verlag.

Selander, R. K. 1976 Genic variation in natural populations. In *Molecular evolution* (ed. F. J. Ayala), pp. 21–45. Sunderland, Mass.: Sinauer Associates.

Sheppard, P. M. 1952 A note on non-random mating in the moth *Panaxia dominula* (L.). *Heredity, Lond.* **6**, 239–241.

Sheppard, P. M. 1959 The evolution of mimicry: a problem in ecology and genetics. *Cold Spring Harb. Symp. quant. Biol.* **24**, 131–140.

Sheppard, P. M. 1975 *Natural selection and heredity*, 4th edn. London: Hutchinson University Library.

Sheppard, P. M. & Cook, L. M. 1962 The manifold effects of the medionigra gene in the moth *Panaxia dominula* and the maintenance of a polymorphism. *Heredity, Lond.* **17**, 415–426.

Sim, G. K., Efstratiadis, A., Jones, C. W., Kafatos, F. C., Koehler, M., Kronenberg, H. M., Maniatis, T., Regier, J. C., Roberts, B. F. & Rosenthal, N. 1978 Studies on the structure of genes expressed during development. *Cold Spring Harb. Symp. quant. Biol.* **42**, 933–945.

Singh, R., Lewontin, R. C. & Felton, A. 1976 Genetic heterogeneity within electrophoretic 'alleles' of xanthine dehydrogenase in *Drosophila pseudoobscura*. *Genetics, Princeton* **84**, 609–629.

Sinnock, P. 1970 Frequency dependence and mating behaviour in *Tribolium casteneum*. *Am. Nat.* **104**, 469–476.

Soane, I. D. & Clarke, B. 1973 Evidence for apostatic selection by predators using olfactory cues. *Nature, Lond.* **241**, 62–64.

Spiess, E. B. 1968 Low frequency advantage in mating of *Drosophila pseudoobscura* karyotypes. *Am. Nat.* **102**, 363–379.

Spofford, J. B. 1969 Heterosis and the evolution of duplications. *Am. Nat.* **103**, 407–432.
Sved, J. A., Reed, T. E. & Bodmer, W. F. 1967 The number of balanced polymorphisms that can be maintained in a natural population. *Genetics, Princeton* **55**, 469–481.
Thoday, J. M. 1972 Disruptive selection. *Proc. R. Soc. Lond.* B **182**, 109–143.
Turner, J. R. G. 1977 Butterfly mimicry: the genetical evolution of an adaptation. *Evolut. Biol.* **10**, 163–206.
van Delden, W., Boerema, A. C. & Kamping, A. 1978 The alcohol dehydrogenase polymorphism in populations of *Drosophila melanogaster*. I. Selection in different environments. *Genetics, Princeton* **90**, 162–191.
Vane-Wright, R. I. 1976 A unified classification of mimetic resemblances. *Biol. J. Linn. Soc.* **8**, 25–56.
Vigue, C. L. & Johnson, F. M. 1973 Isozyme variability in species of the genus *Drosophila*. VI. Frequency–property–environment relationships of allelic alcohol dehydrogenases in *D. melanogaster*. *Biochem. Genet.* **9**, 213–227.
Vogel, F. 1975 ABO blood groups, the HLA system and diseases. In *The role of natural selection in human evolution* (ed. F. M. Salzano), pp. 247–269. Amsterdam: North-Holland.
Wallace, B. 1970 *Genetic load: its biological and conceptual aspects*. Englewood Cliffs, N.J.: Prentice-Hall.
Wallace, B. 1975 Hard and soft selection revisited. *Evolution* **29**, 465–473.
Ward, R. D. 1977 Relationship between enzyme heterozygosity and quaternary structure. *Biochem. Genet.* **15**, 123–135.
Wickler, W. 1968 *Mimicry in plants and animals*. London: Weidenfeld & Nicholson, World University Library.
Williamson, M. H. 1957 An elementary theory of interspecific competition. *Nature, Lond.* **180**, 422–425.
Wills, C., Phelps, J. & Ferguson, R. 1975 Further evidence for selective differences between isoalleles in *Drosophila*. *Genetics, Princeton* **79**, 127–141.
Wilson, A. C. 1976 Gene regulation in evolution. In *Molecular evolution* (ed. F. J. Ayala), pp. 225–234. Sunderland, Mass.: Sinauer Associates.
Wright, S. 1969 *Evolution and the genetics of populations*, vol. 2: *The theory of gene frequencies*. Chicago: University of Chicago Press.
Yamazaki, T. 1971 Measurement of fitness at the esterase-5 locus of *Drosophila pseudoobscura*. *Genetics, Princeton* **67**, 579–603.
Yamazaki, T. 1977 Enzyme polymorphism and functional difference: mean, variance and distribution of heterozygosity. In *Molecular evolution and polymorphism* (ed. M. Kimura), pp. 127–147. Mishima, Japan: National Institute of Genetics.
Yu, P. 1972 Some host–parasite interaction models. *Theor. Popul. Biol.* **3**, 347–357.

Proc. R. Soc. Lond. B **205**, 475–488 (1979)
Printed in Great Britain

Game theory and the evolution of behaviour

By J. Maynard Smith, F.R.S.
School of Biological Sciences, University of Sussex, Falmer, Brighton BN1 9QG, U.K.

How far can game theory account for the evolution of contest behaviour in animals? The first qualitative prediction of the theory was that symmetric contests in which escalation is expensive should lead to mixed strategies. As yet it is hard to say how far this is borne out, because of the difficulty of distinguishing a 'mixed evolutionarily stable strategy' maintained by frequency-dependent selection from a 'pure conditional strategy'; the distinction is discussed in relation to several field studies. The second prediction was that if a contest is asymmetric (e.g. in ownership) then the asymmetry will be used as a conventional cue to settle it. This prediction has been well supported by observation. A third important issue is whether or not information about intentions is exchanged during contests. The significance of 'assessment' strategies is discussed.

Introduction

The theory of games was applied to the evolution of fighting behaviour (Maynard Smith & Price 1973) in the hope of explaining the conventional aspects of this behaviour (Lorenz 1966 and earlier) in terms of individual selection. Before this there had been a tendency to account for such behaviour in terms of the good of the species. This presented an obvious difficulty to a neo-Darwinist, in view of the weakness of group compared with individual selection as an evolutionary force (Williams 1966; Maynard Smith 1964, 1976).

The aim of this paper is to discuss how far the application of game theory, and in particular the concept of an 'evolutionarily stable strategy' is useful in interpreting animal contests. By 'interpreting' I mean giving an account of the selective forces responsible for the evolution of fighting behaviour: the short-term developmental and physiological mechanisms call for a different type of explanation. That is, I am seeking a 'functional' rather than a 'causal' explanation. I am not concerned to show that any particular model – for example the Hawk–Dove game – is an adequate account of any particular case, because clearly it is not and was not intended to be. Rather, I ask whether certain general predictions of the game theory approach are borne out, and whether there are categories of observation that cannot at present be understood.

The predictions of game theory

The following three general predictions follow from game theory models of evolution:

(i) In symmetric contests, we expect to find 'mixed' strategies.

(ii) In asymmetric contests, we expect the asymmetry to be used as a cue to settle the contest.

(iii) We do not expect contestants to exchange information about their intentions, although they certainly exchange information about their size and fighting ability.

Table 1. Payoff matrix for contests between Hawk (H) and Dove (D)

(The payoffs are to the player adopting the strategy on the left against one adopting the strategy above.)

	H	D
H	−5	+10
D	0	+5

These predictions are now derived and explained in terms of a simple model. Suppose a contest takes place between two individuals over some resource. Individuals can adopt two 'strategies': H (Hawk), i.e. 'fight until victory or defeat', and D (Dove), i.e. 'display, but retreat before being hurt if one's opponent escalates'. The changes in fitness after a contest can then be expressed in a 'payoff matrix', which might be as in table 1. In effect, what this matrix says is that if both contestants adopt H, their expected losses from injury or exhaustion more than outweigh their expected gains; that H always wins against D, but D loses nothing because he retreats before injury; and that D sometimes wins and sometimes loses against another D.

Now imagine a population of individuals playing such a game, and then producing offspring in numbers proportional to their accumulated payoffs, the offspring adopting the same strategy as their parents. The state (or states) to which such a population evolves is called an 'evolutionarily stable strategy' (e.s.s.). In words, to be an e.s.s. a strategy I must have the following property: if almost all the members of a population adopt I, then no mutant strategy M can invade the population. This in turn requires that, for all possible mutants, the expected payoff to M against I must be less than (or at least not greater than) the payoff to I against itself. A mathematical definition is given by Maynard Smith (1974).

With this idea in mind, it is clear that neither H nor D is an e.s.s. for the game in table 1. Thus H is not an e.s.s. because, in a population of Hawks, the expected payoff to H per contest is −5, and to a D mutant is 0, which is greater. A similar argument shows that D is not an e.s.s. What then would happen to a population containing both H and D strategists? It can be shown that it would evolve to a stable mixture of H and D (for the particular numerical values, H and D would be equally frequent). Alternatively, if the strategy *I*, 'play H with probability $\frac{1}{2}$, and D with probability $\frac{1}{2}$', could exist and reproduce, then *I* would be an e.s.s.; such a strategy, containing a stochastic element, is called a 'mixed' strategy.

Table 2. Payoff matrix for contests between Hawk (H), Dove (D) and Bourgeois (B)

	H	D	B
H	−5	+10	+2.5
D	0	+5	+2.5
B	−2.5	+7.5	+5

This, then, is our first prediction, that symmetric games tend to have mixed e.s.ss. The exact numbers do not matter, but it is important that the cost of injury outweighed the expected gains of victory. It is an open question whether, in such cases, we should expect a genetically polymorphic population, or a uniform population of phenotypically flexible mixed strategists.

Now suppose that each contest takes place between the 'owner' of a resource and an 'interloper'. A new strategy now becomes possible: *B* (Bourgeois), or 'play H if owner, D if interloper'. The new payoff matrix is shown in table 2. The essential point here is that an escalated fight never takes place between two *B* strategists, because if one is owner the other is not. Clearly, the strategy *B* is an e.s.s. for the matrix; in fact it is the only e.s.s. This is our second prediction, that in asymmetric contests the asymmetry will be used as a cue. The asymmetry need not be of ownership. It could as well be a difference in size or weapons, provided two conditions are satisfied. First, the asymmetry must be unambiguously perceived by both contestants. Secondly, it must not be easy to acquire the 'winning' cue; if it were, every member of the population would acquire it and the asymmetry would disappear.

The third prediction, that information about intentions should not be conveyed, will be explained later when the relevant data are discussed.

Are there mixed e.s.ss?

Contest behaviour is often very variable, but this does not prove that we are looking at a mixed e.s.s. The difficulty can best be illustrated by an example. Rohwer (1977) studied the behaviour of Harris sparrows in winter flocks. These flocks are temporary, so birds probably do not recognize one another individually. Birds differ strikingly in the amount of black in their plumage. Black birds are more aggressive, and regularly occupy higher positions in the flock hierarchy. These observations raise two questions. The first, relevant here, is why there are individual differences in aggression. The second, relevant to the problem of information transfer, is why these differences are signalled by differences in plumage.

To the first question there are two possible answers. First, the population may be at a mixed e.s.s. Secondly, some sparrows are large and some small (perhaps for chance environmental reasons), and it would not pay a small sparrow to be aggressive. If the latter answer is true, all sparrows adopt the same pure, conditional, strategy: 'if large, play H; if small, play D'; in effect, when small the animal makes the best of a bad job. As others have emphasized (e.g. Parker 1978) there is a crucial difference between a mixed e.s.s. and a pure conditional strategy. But how are they to be distinguished in practice?

I suggest that, to be interpreted as a mixed e.s.s., a strategy must meet two criteria. First, the fitness, over a lifetime, of individuals adopting different strategies must be the same (or, for a true mixed strategist, the payoffs when adopting different policies should be the same). This is a theoretical requirement (Bishop & Cannings 1978) which is easy to see intuitively. Thus suppose mixed strategy I plays sometimes H and sometimes D, and does better on the occasions it plays H; clearly, a strategy J, playing H more often and D less, would do better, so I is not an e.s.s. The second criterion is that if any component of the supposed mixed e.s.s. became more frequent it would have a lower payoff; that is, fitnesses are negatively frequency-dependent. Although hard to prove in the field, it is often easy to see that this would be the case.

Not surprisingly, there are few cases which meet these criteria. Parker (1974*a*) showed that they hold for male dung flies, which have the choice between sticking with a stale cowpat despite the declining rate at which virgin females arrive, or seeking a fresh one. The Harris sparrow may also be an example of a mixed e.s.s., although the situation is not simple. It is difficult to measure individual survival, but the frequencies of light and dark birds are fairly constant through the winter, which fits the e.s.s. hypothesis (S. Rohwer, personal communication). However, there is a rise in the status signalled by individuals between the first and second winters, which makes more sense if it pays to be dominant. Rohwer & Rohwer (1978) have proposed a frequency-dependent mechanism to account for the polymorphism.

More recently, Brockmann *et al.* (1979) have suggested that the criteria are met by the digger wasp, *Sphex ichneumoneus*. Females of this species dig and provision burrows for their eggs. An individual has a choice of digging her own burrow, or entering a burrow already dug by another wasp. The burrow may have been abandoned, in which case entering it is a good policy. Alternatively, the 'owner' may be away hunting. If so, she may find the interloper when she returns, and there is a fight; in these cases 'entering' is a less satisfactory strategy. In one population studied, the average payoffs to the strategies 'enter' and 'dig' were indeed equal. Individual females adopt sometimes one strategy and sometimes the other; they are mixed strategists. It is easy to see that the payoffs would be frequency-dependent in the required way. The only snag is that, although the data are beautifully in accord with the mixed e.s.s. hypothesis in one population (in Maine), they are not so in a population in Michigan.

There are many cases which it is tempting to interpret as mixed e.s.ss, but in which crucial data are missing. It is common to find variability in male mating strategies; e.g. client males in the ruff (Hogan-Warburg 1966; Van Rhijn 1973) or sea-going and precociously mature salmon (Jones 1969). Gadgil (1972) suggested that some such cases may be stable genetic polymorphisms. Others are obviously examples of a pure conditional strategy. For example, in the red deer the strategy 'when adult, hold a harem; when young, try to abduct a hind' is clearly a pure conditional strategy.

In the tree frog, *Hyla cinerea*, most males call, but some satellite males stay close to a calling male and try to intercept and mate with the females he attracts (Perrill *et al.* 1978). Some males employed the satellite strategy from night to night; others changed strategies from night to night, or on the same night. There is no consistent size difference between calling and satellite males. Thirty experiments in which a gravid female was released close to a pair of males resulted in the calling male achieving amplexus on 17 occasions and the satellite on 13. These facts are at least consistent with a mixed e.s.s. interpretation.

An entertaining example is provided by the study by Alcock *et al.* (1977) of the bee *Centris pallida*. Male bees adopt one of two strategies. 'Patrollers' search for sites where buried virgin females are about to emerge, dig for them and attempt mating. Fights over digging sites are common, and are usually won by the larger male. 'Hoverers' wait for females that have not been mated by patrollers, perhaps because a fight was in progress when they emerged. Hoverers are smaller than patrollers. Although measurements of mating success are not available, it seems certain that the larger patrollers are more fit. To a small male, hovering is making the best of a bad job. But the size of a male is determined by the amount of food provided by his mother. Small males are cheaper to produce, both because they need less food and because they are less likely to be parasitized. It is hard to see how producing small males could be forced on a female by circumstances. A possible explanation, pointed out to me by Dr Jane Brockmann, is that females are adopting a mixed e.s.s. and thereby forcing a pure conditional strategy on their sons.

It is early to say how often variation in contest behaviour is properly to be regarded as a mixed e.s.s. At this stage, the important point is that, in a mixed e.s.s., the payoffs to the different components of the strategy should be equal.

Are asymmetries taken as clues?

The most contra-intuitive prediction of game theory is that 'ownership' may be taken as a cue to settle contests, even if it does not affect the likelihood of winning an escalated fight or the payoff for winning. To demonstrate that ownership can settle contests irrespective of differences in r.h.p. ('resource-holding power'; Parker 1974*b*), it is sufficient to show three things: (i) contests are won by owners; (ii) a contest between two specific individuals may come out sometimes one way and sometimes the other, according to who is the owner; (iii) if two contestants can both be persuaded that they are the owner of the same resource, an escalated contest ensues of a kind not normally observed between owner and interloper.

It is not sufficient merely to establish (i), because that would be true if only strong individuals became owners in the first place. All these points were established by Kalmus (1941) for contests between groups of worker bees over food sources; more recently, they have been shown for contests over sun spots (mating territories) in the speckled wood butterfly, *Parage aegeria* (Davies 1978), and for contests over females in the Hamadryas baboon (Kummer 1971). Points (i) and (iii) were established for contests over hilltops (mating territories) in the butterfly *Papilio zelicaon* (L. E. Gilbert, personal communication). Baker (1972) showed that in the peacock butterfly, *Inachis io*, territorial contests are normally won by the owner after a spiral flight. If a second male settles in a territory (e.g. because the owner is absent courting), the contest is still usually settled in favour of the owner, probably because the interloper cannot find his way back to the territory after a spiral flight. However, if the interloper overcomes this difficulty, the contest is settled after further flights in favour of the male able to hold his position above and behind his opponent.

Parker (1978) lists some other cases in which there are strong reasons for thinking that ownership settles contests. It is likely that the phenomenon will prove to be general.

It is more readily accepted that asymmetries in size or weapons can settle a contest without escalation. Theoretical analysis (Maynard Smith & Parker 1976) confirms that acceptance of such asymmetries to settle contests can be an e.s.s., provided that the asymmetry is known unambiguously to both contestants. If recognition of the asymmetry is ambiguous then each contestant might perceive that it was superior (e.g. bigger), leading to an escalated contest. The asymmetry may be unambiguously recognized but a poor guide to the outcome of an escalated fight; in this case, the asymmetry may still be effective in settling contests provided the risks of escalation are high.

To give a recent example, Jakobsson *et al.* (1979) studied fighting in the cichlid fish *Nannacara anomala*. Pairs of males were matched carefully for size, a pair differing by less than 0.2 g in a total of 5 g, yet the heavier fish won in 10 out of 11 fights (Jacobsson, personal communication). The mouth fighting, which is a striking feature of their behaviour, may enable even small differences of weight to be detected, as was suggested for another cichlid by Apfelbach & Leong (1970). It is puzzling, however, that in other species size appears to have little effect on ultimate outcome. For example, Dow *et al.* (1976) claim that there is no correlation between size, coloration or fin area and winning in the fish *Aphysemion striatum* or *A. bivittatum* (although no data are given).

Real difficulties arise when there are asymmetries both of size and ownership in the same contest. Hyatt & Salmon (1978) observed 403 fights in the field between male fiddler crabs, *Uca pugilator*, over burrows. There was no average size difference between owners and interlopers. The owner won in 349 cases and the interloper in 54. However, in these 54 cases the interloper was larger in 50 cases and smaller only in 1. It seems that ownership is typically taken as the arbiter, but that a sufficiently large size difference can override this. The males have claws strong enough to crush the abdomen of an opponent, but injury has never been observed in these fights.

A different balance is struck in the funnel-web spider, *Agelenopsis aperta* (Riechert 1978). Fights between females over webs were observed in the field. In 81 fights the larger female won in all but 7 cases. However, in 10 cases in which the size difference was less than 10% the owner won on 9 occasions. Thus size is taken as the primary cue, but ownership decides if the size difference is ambiguous. It is worth commenting that shaking a web gives rather reliable information about relative mass.

An even more difficult case is the study by Rand & Rand (1976) of contests between gravid *Iguana* females over egg-laying burrows; 80% of these fights are won by owners. Here, 'owner' means the female temporarily in possession of the burrow. However, since a female cannot build a burrow in a single shift but must take numerous rests, the 'interloper' may be the female who has done most of the digging on the burrow, and hence in a different sense may be the owner. Data on this point are not available, but can hardly be irrelevant. Unless previous work on a burrow in some degree establishes ownership, it is hard to see why a lizard would ever *start* digging a burrow, since only the lizard which finishes a burrow gains any payoff. (In the similar case, discussed above, of the wasp *Sphex ichneumoneus*, the winner of a fight over a burrow is usually the female that has provided the larger number of katydids (Dawkins & Brockman 1979). The length of fights correlates closely with the number of katydids provided by the loser.)

Some theoretical work has been done on contests with both ownership and size (r.h.p.) asymmetries. P. Hammerstein has considered a model in which size asymmetry, if present, can be assessed unambiguously and without cost. He considers two strategies, 'Bourgeois', which ignores r.h.p. asymmetry and considers only ownership, and 'Assessor', which plays H if larger than its opponent, D if smaller, and Bourgeois if there is no detectable size difference (*Agelenopsis* females, described above, adopt the Assessor strategy). If escalation is dangerous, the Assessor strategy is always an e.s.s. The Bourgeois strategy may also be an e.s.s., so that the population may have two alternative stable states.

Is information transferred during contests?

This question, and the related ones of bluff and lying, are the main problem areas for a game theoretical analysis of contest behaviour. Maynard Smith & Parker (1976) argued that animals would not convey information about their intentions on the following grounds. Suppose that a population existed in which individuals conveyed accurate information about their 'intentions' – i.e. about the level to which they would escalate. If an individual found that an opponent was announcing a higher intention than its own, it would pay to retreat at once. Therefore a 'lying' mutant which announced a very high (but untrue) intention would be favoured by selection. Before long, everyone would be lying, and it would then pay to ignore the message being transmitted. The argument was put in more mathematical language, but it simply amounts to saying that you should not believe what an opponent at poker tells you.

The difficulty, of course, is that an extensive literature exists to demonstrate that, in some sense, information *is* transmitted during contests (see, for example, Simpson (1968) and Dow *et al.* (1976) on fish; Hazlett (1966, 1972), Hazlett & Bossert (1965) and Dingle (1969) on crustacea; Stokes (1962*a*, *b*), Dunham (1966) and Andersson (1976) on birds; and Rand & Rand (1976) on lizards). What these papers show is that (i) in the course of a contest animals use not one kind of act but a range of acts indicating different levels of aggression; (ii) an act contains information about the future intentions of the actor, since it is correlated with his future acts; and (iii) the information is transmitted, since the acts of the receiver are influenced by the actor. How is this apparent discrepancy between theory and observation to be reconciled?

No complete answer to this question can yet be given, because the data were often collected with no very clear functional question in mind. Some points are, however, worth making:

(i) Much of the information transmitted concerns 'resource-holding potential' (r.h.p.) (Parker 1974*b*) and not intentions. Briefly, r.h.p. affects the outcome of an escalated contest, and cannot be increased without cost in time or resources; intentions do not affect the outcome of an escalated contest, and can be altered without cost (except via their consequences). It is essential to distinguish between the messages 'I am 6 ft tall' and 'I am going to hit you', because the former cannot easily be a lie and the latter may be (this is not intended to detract from the importance of ruffs, manes, markings, etc. whose function is to make an animal look as large as possible). As stated earlier, an asymmetry in size can be used to settle a contest. If the asymmetry cannot be immediately perceived, an assessment phase of the contest can also be part of an e.s.s.; this holds even if assessment costs something, provided it is cheaper than escalation and yields a reasonably unambiguous result.

No doubt much of the behaviour described by Riechert (1978) in spiders and by Hyatt & Salmon (1978) in fiddler crabs is concerned with r.h.p. assessment, because they have also shown that size differences do influence outcome.

The information used in assessment should be difficult to fake: that is, it should be impossible for an individual by a small expenditure of resources to signal a high r.h.p. In American bighorn sheep, Geist (1966) has shown that differences in horn size can settle contests without a fight. Horns are expensive; they can amount to 10% of body mass. Geist makes the interesting suggestion that the great variability of horn size in these sheep is related to their function in assessment. In red deer, most fights are preceded by roaring contests. Roaring rate in mature stags correlates with fighting success, and many fights are settled by roaring, without physical contact. Clutton-Brock & Albon (1979), who report these facts, point out that since fighting ability declines as an individual becomes exhausted during the rut, roaring rate may be a more reliable assessor than mere size. Further, roaring rate, like fighting ability, declines with advancing age, whereas size does not. To give one more example of an assessment which is difficult to fake, Davies & Halliday (1978) show that in toads the depth of pitch of calls is related to body size, and is used to decide whether to escalate contests over females.

(ii) Little information about long-term intentions is in fact transmitted. Thus an act may be correlated with an immediately preceding or following act, and yet be a poor guide to whether a contestant will ultimately retreat or attack. Caryl (1979) has reanalysed the data on tits (Stokes 1962*a*, *b*), grosbeaks (Dunham 1966) and skuas (Andersson 1976) and concludes that the displays are a poor predictor of future attack (although some displays are a good predictor of immediate retreat; I shall return to this point). Thus no one display is followed by attack with high probability; a particular display may be correlated with attack at one time of year but not another; in two of three species, the most aggressive display by one contestant did not correlate with retreat by the other; two different displays may have the same effect on the receiver but predict different things about the actor.

None of this makes much sense if one thinks that the function of these displays is to transmit information about intentions. In fish there is again little reason to think that intentions are signalled. Simpson (1968), in his study of *Betta splendens*, starts by saying that 'a criterion for a successful description of the threat display is the prediction of the outcome of encounters from the differences between the displays of the participants'. He finds one feature, the proportion of time during which the gill covers are raised, which does predict the outcome, but only during the last 2 min of a fight. Earlier in the same fights, winners exceeded losers in this respect in four cases, losers exceeded winners in two cases, and in two cases there was no difference; hardly a successful prediction.

Dow *et al.* (1976) studied the fish *Aphysemion striatum*. The most aggressive act (biting the opponent) was commoner in eventual winners in the *first* quarter of the fight, but there was no difference between contestants subsequently. Now it is inconceivable that it could be selectively advantageous for a fish to tell its opponent at the start of a fight that it intended to retreat, and then to continue the fight; it is as if a trade union negotiator were to say 'I demand a 50% increase for my members; furthermore, I will settle for 5%'. The explanation must lie in the artificiality of the experimental procedure (the fight was initiated by the sudden removal of an opaque barrier separating the fish), a point which is recognized by the authors.

Jakobsson *et al.* (1979) observed fights in the cichlid fish *Nannacara anomala*, also in the laboratory. Both the frequencies of acts, and the reaction pattern to each act from the other fish, were very similar in ultimate winners and losers, differences appearing mainly towards the end of a fight. However, after receiving a bite (the only act likely to cause injury) a fish often retaliated by biting back. This illustrates the point that there may be a correlation between successive acts, so that 'information' is in a sense being transferred, yet an act may convey little information about long-term intentions to retreat or continue. As mentioned earlier, these fights are probably settled by size assessment during mouth wrestling.

(iii) 'Lying' may be impossible, or may be punished. The theoretical case against information transfer rested essentially on the argument that accurate information transfer would not be proof against invasion by a lying mutant. The concept of lying in an animal requires more careful definition. If in most members of a species the signal A is usually followed by the act X, then the signal A may be said to 'mean' X, and a mutant who performs signal A but *not* act X may be said to be lying. Rohwer's (1977) experiments on Harris sparrows are suggestive here. Typically, dark birds are aggressive and pale ones timid. To be dark can be interpreted as a signal, because all birds are capable of developing dark plumage, and do so in the spring. By painting dark birds light, and light ones dark, Rohwer constructed artificial 'liars'. These liars, of both kinds, received more than their fair share of attacks from other birds. More recently (Rohwer, personal communication) it has been shown that pale birds injected with testosterone were also persecuted, but if at the same time they were painted dark (bringing behaviour and signal into line) they could succeed as dominants.

Now if it were part of the behavioural strategy of normal individuals to 'punish' lying, this could stabilize accurate information transfer, rather as 'retaliation' can stabilize conventional as opposed to escalated behaviour. The difficulty, of course, is to explain why 'punishing' should be typical behaviour; it is easy to see that punishing would be good for the species, but less clear what could be the advantage to the individual.

(iv) The advantage of a 'surrender' signal. The most striking feature of Caryl's reanalysis of bird contest data was to show that although there are no good predictors of attack, there *are* good predictors of retreat – for example, erecting the crest in the blue tit. Similarly Dow *et al.* (1976) found in *Aphysemion* that a particular act, 'fin clamp', although generally rare, was the last act of the loser in 18 out of 21 fights. This was not a prediction of any game theory model, but it does make sense. It pays the sender to give the signal, because it prevents a damaging attack by the winner; it pays the winner to observe it (i.e. not to launch a damaging attack), because an attack would gain nothing and would risk retaliation; there would be no point in lying, because a bird which signals 'surrender' and then does not is no better off than a bird which gave no signal.

(v) Communication is useful if resources are divisible. The obvious divisible resource is space, e.g. in territorial behaviour. There are theoretical grounds for thinking that in a contest over a divisible resource, communication by graded signals could evolve. There is need for further work on this possibility.

Discussion

Evolutionary game theory is, I believe, the appropriate framework for a functional interpretation of contest behaviour. There is no shortage of problems, but progress has been rapid. There are, however, some general criticisms of the approach that should be discussed.

First, there has been criticism of the basic model itself. The model assumes parthenogenetic inheritance; that is, it assumes that like begets like but does not allow for the complexities of sexual reproduction. Auslander *et al.* (1978) describe a two-species system with sexual reproduction and density-dependent population regulation, and show that although a static e.s.s. exists the populations may not reach this e.s.s. when genetic and population dynamics are included. Their results are of considerable interest, but do not seriously undermine the use of e.s.s. theory for the analysis of phenotypic evolution in single species. Further justification for the belief that, given sufficient genetic flexibility, selection will equalize fitnesses of different phenotypes at an e.s.s. is provided by Slatkin (1978).

The e.s.s. model also assumes an infinite and randomly mixing population. Riley (1978) has shown that the stability criteria for a finite population are different. Both criticisms are formally correct. It is certainly possible to construct cases in which sex or finite population size will alter the outcome. This should be borne in mind, but to introduce these complexities into the model at this stage would in my view be a mistake. It would be to abandon the simplicity and predictive power of the model in a fruitless pursuit of precision. There is, however, one extension of the basic model which is worth pursuing. Actual populations are viscous, so that relatives live close to one another. Contests are therefore often between relatives. Extensions of the model to allow for this have been proposed by Mirmirani & Oster (1978) and Treisman (1977), and, from a different standpoint, by Grafen (1979) and Hines & Maynard Smith (1979).

From an observational standpoint, the main criticism has been by Geist (1974). His main complaints were that we wrongly supposed that there is a distinction between 'dangerous' and 'conventional' behaviour in animal contests, and that we failed to do justice to earlier explanations of contest behaviour in terms of individual selection. There is, I think, some justice in these complaints. Geist is surely correct in emphasizing that animals do injure and even kill one another, and that their behaviour and weapons can be interpreted in terms of efficiency in defence and attack (Geist 1966). In fact, we were not concerned to deny this; indeed, our interpretations would be meaningless unless escalation and injury do sometimes occur. Nevertheless, there are phenomena, such as the acceptance of ownership discussed in § 4, or the 'deliberate' aiming at the best-protected part of an opponent, (see, for example, Dingle & Caldwell (1969) on stomatopods) which can reasonably be called 'conventional' and which demand explanation.

In fact, the disagreement between Geist (1974) and Maynard Smith & Price (1973) was more apparent than real. Part of the difficulty arose because of our failure to notice that Geist had earlier put forward part of our argument. In addition to attempting a general formulation for the analysis of contests, we proposed that a specific strategy, namely 'retaliation', was likely to be a component of evolutionarily stable behaviour. I think that this conclusion is correct (despite some mathematical difficulties; see Gale & Eaves 1975), but it had been stated quite explicitly by Geist (1966). In a later analysis of contest behaviour in ungulates, V. Geist has given the concept of retaliation a more central position and has extended his earlier argument in an attempt to relate the types of weapons and behaviour to the ecology of different species. Had there been space, the role of retaliation would have been treated here as a fourth prediction of game theory; it was omitted only because it is observationally less controversial.

References (Maynard Smith)

Alcock, J., Jones, C. E. & Buchmann, S. L. 1977 Male nesting strategies in the bee *Centris pallida* Fox (Anthophoridae: Hymenoptera). *Am. Nat.* **111**, 145–155.

Andersson, M. 1976 Social behaviour and communication in the Great Skua. *Behaviour* **58**, 40–77.

Apfelbach, R. & Leong, D. 1970 Zum Kampfverhalten in der Gattung *Tilapia* (Pisces, Cichlidae). *Z. Tierpsychol.* **27**, 98–107.

Auslander, D., Guckenheimer, J. & Oster, G. 1978 Random evolutionarily stable strategies. *Theor. Popul. Biol.* **13**, 276–293.

Baker, R. R. 1972 Territorial behaviour of the nymphalid butterflies, *Aglais urticae* (L.) and *Inachis io* (L.) *J. Anim. Ecol.* **41**, 453–469.

Bishop, D. T. & Cannings, C. 1978 A generalised war of attrition. *J. theor. Biol.* **70**, 85–124.

Brockmann, H. J., Grafen, A. & Dawkins, R. 1979 Evolutionarily stable nesting strategy in a digger wasp. *J. theor. Biol.* (In the press.)

Caryl, P. G. 1979 Communication by agnostic displays: what can games theory contribute to ethology? *Behaviour.* (In the press.)

Clutton-Brock, T. H. & Albon, S. D. 1979 The roaring of red deer and the evolution of honest advertisement. *Behaviour.* (In the press.)

Davies, N B. 1978 Territorial defence in the speckled wood butterfly (Pararge aegeria): the resident always wins. *Anim. Behav.* **26**, 138–147.

Davies, N. B & Halliday, T. M. 1978 Deep croaks and fighting assessment in toads *Bufo bufo*. *Nature, Lond.* **274**, 683–685.

Dawkins, R. & Brockman, H. J. 1979 Do digger wasps commit the concorde fallacy? *Anim. Behav.* (In the press.)

Dingle, H. 1969 A statistical and information analysis of aggressive communication in the mantis shrimp. *Gonodactylus bredini* Manning. *Anim. Behav.* **17**, 561–575.

Dingle H. & Caldwell, R. L. 1969 The aggressive and territorial behaviour of the mantis shrimp *Gonodactylus bredini* Manning (Crustacea, Stomatopoda). *Behaviour* **33**, 115–136.

Dow, M., Ewing, A. W. & Sutherland, I. 1976 Studies on the behaviour of cyprinodont fish. III. The temporal patterning of aggression in *Aphysemion striatum* (Boulenger). *Behaviour* **59**, 252–268.

Dunham, D. W. 1966 Agnostic behaviour in captive Rose-breasted Grosbeaks (*Phenoticus ludovicianus*) (L.) *Behaviour* **27**, 160–173.

Gadgil, M. 1972 Male dimorphism as a consequence of sexual selection. *Am. Nat.* **106**, 574–580.

Gale, J. S. & Eaves, L. J. 1975 Logic of animal conflict. *Nature, Lond.* **254**, 463–464.

Geist, V. 1966 The evolution of horn-like organs. *Behaviour* **27**, 175–213.

Geist, V. 1974 On fighting strategies in animal conflict. *Nature, Lond.* **250**, 354.

Grafen, A. 1979 The problem of games between relatives. *Anim. Behav.* (In the press.)

Hazlett, B. 1966 Factors affecting the aggressive behaviour of the hermit crab *Calcinus tibicen*. *Z. Tierpsychol.* **23**, 655–671.

Hazlett, B. 1972 Stimulus characteristics of an agnostic display of the hermit crab (*Calcinus tibicen*). *Anim. Behv.* **20**, 101–107.

Hazlett, B. & Bossert, W. H. 1965 A statistical analysis of the aggressive communications systems of some hermit crabs. *Anim. Behav.* **13**, 357–373.

Hines, W. G. S. & Maynard Smith, J. 1979 Games between relatives. *J. theor. Biol.* (In the press.)

Hogan-Warburg, A. J. 1966 Social behaviour of the Ruff, *Philomachus pugnax* (L.). *Ardea* **54**, 109–229.

Hyatt, G. W. & Salmon, M. 1978 Combat in the fiddler crabs *Uca pugilator* and *U. pugnax*: a quantitative analysis. *Behaviour* **65**, 182–211.

Jacobsson, S., Radesäter, T. & Järvi, T. 1979 On the fighting behaviour of *Nannacara anomala* (Pisces, Cichlidae) ♂♂. *Z. Tierpsychol.* (In the press.)

Jones, J. W. 1969 *The salmon*. London: Collins.

Kalmus, H. 1941 Defence of source of food by bees. *Nature, Lond.* **148**, 228.

Kummer, H. 1971 *Primate societies*. Chicago: Aldine-Atherton.

Lorenz, K. 1966 *On aggression*. London: Methuen.

Maynard Smith, J. 1964 Group selection and kin selection. *Nature, Lond.* **201**, 1145–1147.

Maynard Smith, J. 1974 The theory of games and the evolution of animal conflicts. *J. theor. Biol.* **47**, 209–221.

Maynard Smith, J. 1976 Group selection. *Q. Rev. Biol.* **51**, 277–283.

Maynard Smith, J. & Parker, G. A. 1976 *Anim. Behav.* **24**, 159–175.

Maynard Smith, J. & Price, G. R. 1973 The logic of animal conflict. *Nature, Lond.* **246**, 15–18.

Mirmirani, M. & Oster, G. 1978 Competition, kin selection and evolutionary stable strategies. *Theor. Popul. Biol.* **13**, 304–339.

Parker, G. A. 1974*a* The reproductive behaviour and the nature of sexual selection in *Scatophaga stercoraria* L. IX. Spatial distribution of fertilisation rates and evolution of male search strategy within the reproductive area. *Evolution* **28**, 93–108.

Parker, G. A. 1974*b* Assessment strategy and the evolution of animal conflicts. *J. theor. Biol.* **47**, 223–243.

Parker, G. A. 1978 Evolution of competitive mate searching. *A. Rev. Ent.* **23**, 173–196.

Perrill, S. A., Gerhardt, H. C. & Daniel, R. 1978 Sexual parasitism in the green tree frog (*Hyla cinerea*). *Science, N.Y.* **200**, 1179–1180.

Rand, W. M. & Rand, A. S. 1976 Agonistic behaviour in nesting iguanas: a stochastic analysis of dispute settlement dominated by the minimization of energy costs. *Z. Tierpsychol.* **40**, 279–299.

Riechert, S. E. 1978 Games spiders play: behavioural variability in territorial disputes. *Behav. Ecol. Sociobiol.* **3**, 135–162.

Riley, J. G. 1978 Evolutionary equilibrium strategies. *UCLA Working Paper* no. 109, Department of Economics.

Rohwer, S. 1977 Status signaling in Harris sparrows: some experiments in deception. *Behaviour* **61**, 107–129.

Rohwer, S. & Rohwer, F. C. 1978 Status signalling in Harris sparrows: experimental deceptions achieved. *Anim. Behav.* **26**, 1012–1022.

Simpson, M. J. A. 1968 The display of Siamese Fighting Fish, *Betta splendens*. *Anim. Behav. Monogr.* **1**, 1–73.

Slatkin, M. 1978 On the equilibration of fitnesses by natural selection. *Am. Nat.* **112**, 845–859.

Stokes, A. W. 1962*a* Agonistic behaviour among blue tits at a winter feeding station. *Behaviour* **19**, 118–138.

Stokes, A. W. 1962*b* The comparative ethology of Great, Blue, Marsh and Coal Tits at a winter feeding station. *Behaviour* **19**, 208–218.

Treisman, M. 1977 The evolutionary restriction of aggression within a species: a game theory analysis. *J. math. Psychol.* **16**, 167–203.
Van Rhijn, J. G. 1973 Behavioural dimorphism in male ruffs, *Philomachus pugnax* (L.). *Behaviour* **47**, 153–229.
Williams, G. C. 1966 *Adaptation and natural selection*. Princeton University Press.

Proc. R. Soc. Lond. B **205**, 489–511 (1979)
Printed in Great Britain

Arms races between and within species

By R. Dawkins and J. R. Krebs
Department of Zoology, South Parks Road, Oxford OX1 3PS, U.K.

An adaptation in one lineage (e.g. predators) may change the selection pressure on another lineage (e.g. prey), giving rise to a counter-adaptation. If this occurs reciprocally, an unstable runaway escalation or 'arms race' may result. We discuss various factors which might give one side an advantage in an arms race. For example, a lineage under strong selection may out-evolve a weakly selected one ('the life–dinner principle'). We then classify arms races in two independent ways. They may be symmetric or asymmetric, and they may be interspecific or intraspecific. Our example of an asymmetric interspecific arms race is that between brood parasites and their hosts. The arms race concept may help to reduce the mystery of why cuckoo hosts are so good at detecting cuckoo eggs, but so bad at detecting cuckoo nestlings. The evolutionary contest between queen and worker ants over relative parental investment is a good example of an intraspecific asymmetric arms race. Such cases raise special problems because the participants share the same gene pool. Interspecific symmetric arms races are unlikely to be important, because competitors tend to diverge rather than escalate competitive adaptations. Intraspecific symmetric arms races, exemplified by adaptations for male–male competition, may underlie Cope's Rule and even the extinction of lineages. Finally we consider ways in which arms races can end. One lineage may drive the other to extinction; one may reach an optimum, thereby preventing the other from doing so; a particularly interesting possibility, exemplified by flower–bee coevolution, is that both sides may reach a mutual local optimum; lastly, arms races may have no stable end but may cycle continuously. We do not wish necessarily to suggest that all, or even most, evolutionary change results from arms races, but we do suggest that the arms race concept may help to resolve three long-standing questions in evolutionary theory.

> 'Wonderful and admirable as most instincts are, yet they cannot be considered as absolutely perfect: there is a constant struggle going on throughout nature between the instinct of the one to escape its enemy and of the other to secure its prey' (Charles Darwin, in Romanes 1883).

Foxes and rabbits race against each other in two senses. When an individual fox chases an individual rabbit the race occurs on the time scale of behaviour. It is an individual race, like that between a particular submarine and the ship it is trying to sink. But there is another kind of race, on a different time scale. Submarine designers learn from earlier failures. As technology progresses, later submarines are better equipped to detect and sink ships, and later-designed ships are better equipped to resist. This is an 'arms race' and it occurs over a historical time scale. Similarly, over the evolutionary time scale the fox lineage may evolve improved adaptations for catching rabbits, and the rabbit lineage improved adaptations for escaping. Biologists often use the phrase 'arms race' to describe this kind of evolutionary escalation of ever more refined mutual counter-adaptations.

We must not be misled by the language of improvement. As the arms race progresses and predators 'improve', this does not necessarily mean they catch more prey. The prey lineage, after all, is improving too. There seems to be no general reason to expect the average success of animals at out-running or out-witting contemporary enemies, victims, prey or competitors, to improve over evolutionary time. Van Valen (1973) has put this point more generally in his 'Red Queen Hypothesis'. But if modern predators are in general no better at catching modern prey than Eocene predators were at catching Eocene prey, it does at first sight seem to be an expectation of the arms race idea that modern predators might massacre Eocene prey. And Eocene predators chasing modern prey might be in the same position as a Spitfire chasing a jet.

There is some indirect fossil evidence bearing on this. Jerison (1973) measured encephalization quotients of carnivorous and herbivorous mammals from the early Tertiary through to the present. As Gould (1978) summarizes his evidence: 'Both herbivores and carnivores displayed continual increase in brain size during their evolution, but at each stage, the carnivores were always ahead. Animals that make a living by catching rapidly moving prey seem to need bigger brains than plant eaters. And, as the brains of the herbivores grew larger (presumably under intense selection pressure imposed by their carnivorous predators), the carnivores also evolved bigger brains to maintain the differential.' Radinsky (1978) casts doubt on the conclusion that the 'carnivores are always ahead', but his reanalysis of the data does nothing to weaken the conclusion that both carnivores and herbivores grew brainier over the period. He concludes that 'Relative brain size increased independently many times in the evolution of the various orders of mammals. Elucidation of the factors responsible for that major evolutionary trend remains an outstanding problem.' The arms race interpretation favoured by Jerison and Gould seems very plausible to us.

H. B. Cott (1940) made early use of the arms race metaphor to dispose of the then fashionable argument that animal mimicry was too perfect to have been produced by natural selection:

> 'Before asserting that the deceptive appearance of a grasshopper or butterfly is unnecessarily detailed, we must first ascertain what are the powers of perception and discrimination of the insects' natural enemies. Not to do so is like asserting that the armour of a battle-cruiser is too heavy, or the range of her guns too great, without inquiring into the nature and effectiveness of the enemy's armament. The fact is that in the primeval struggle of the jungle, as in

the refinements of civilized warfare, we see in progress a great evolutionary armament race – whose results, for defence, are manifested in such devices as speed, alertness, armour, spinescence, burrowing habits, nocturnal habits, poisonous secretions, nauseous taste, and procryptic, aposematic, and mimetic coloration; and for offence, in such counter-attributes as speed, surprise, ambush, allurement, visual acuity, claws, teeth, stings, poison fangs, and anticryptic and alluring coloration. Just as greater speed in the pursued has developed in relation to increased speed in the pursuer; or defensive armour in relation to aggressive weapons; so the perfection of concealing devices has evolved in response to increased powers of perception...'

Charles Darwin lived before 'arms race' became a fashionable phrase, but he was well aware of the evolutionary phenomenon, and seems to have suspected that it was relevant even to adaptations that appeared to be to the inanimate environment only (Darwin 1859):

'...the structure of every organic being is related, in the most essential yet often hidden manner, to that of all the other organic beings, with which it comes into competition for food and residence, or from which it has to escape, or on which it preys. This is obvious in the structure of the teeth and talons of the tiger; and in that of the legs and claws of the parasite which clings to the hair on the tiger's body. But in the beautifully plumed seed of the dandelion, and in the flattened and fringed legs of the water-beetle, the relation seems at first confined to the elements of air and water. Yet the advantage of plumed seeds no doubt stands in the closest relation to the land being already thickly clothed with other plants; so that the seeds may be widely distributed and fall on unoccupied ground. In the water-beetle, the structure of its legs, so well adapted for diving, allows it to compete with other aquatic insects, to hunt for its own prey, and to escape serving as prey to other animals.'

Although we are sympathetic to this view, it is not our purpose to advocate the universal importance of arms races in all evolutionary change. Ours is the more modest aim to suggest that the arms race metaphor leads to new and productive habits of thought about old and familiar material.

We shall organize our discussion around a two-way classification of arms races. An arms race may be interspecific or intraspecific. Cutting across this classification, it may be symmetric or asymmetric. An asymmetric arms race might be called an 'attack–defence' arms race. It is one in which offensive adaptations on one side are countered by defensive adaptations on the other. Swords get sharper, so shields get thicker, so swords get sharper still. Most of the familiar biological examples are asymmetric, as in the predator–prey or the parasite–host asymmetry. In a symmetric arms race on the other hand, the parties are becoming better and better at doing the same thing. The possession of 2 megaton bombs on one side provokes the manufacture of 3 megaton bombs by the other side and so on. Symmetric arms races are essentially competitive, and biological examples are likely to be found within species, for example in the weaponry of male–male competition for females.

The arms race idea suggests progressive rather than random (Raup 1977) trends in evolution, but we are anxious not to be mistaken as strong partisans for the view that Eldredge & Gould (1972) have dubbed 'phyletic gradualism'. Conscious that modern palaeontology is in a fast moving and excitingly controversial phase (see, for example, any recent number of *Paleobiology*), we are content as ethologists to wait on the touchline. Although for convenience we shall normally use 'gradualistic' language, much of what we say could be rephrased in 'punctuationist' terms (Gould 1977). An exception is our treatment of intraspecific symmetric arms races, which does seem necessarily to imply gradual and continuous change.

In all of this discussion it is important to realize who are the parties that are 'racing' against one another. They are not individuals but lineages. To be sure, it is individuals who attack and defend, individuals who kill or resist being killed. But the arms race takes place on the evolutionary time scale, and individuals do not evolve. It is lineages that evolve, and lineages that exhibit progressive trends in response to the selection pressures set up by the progressive improvements in other lineages. This raises a problem when we consider intraspecific arms races, for instance that between males and females, for males and females are not two independent lineages: males are descended from females and vice versa. But we shall face that problem when we come to it.

Before developing our two-way classification, we need to consider in general those factors that are likely to favour one side or the other in an arms race.

Factors favouring one side or the other in arms races

Slobodkin (1974) wrote that 'while an "evolutionary race" may exist between predators and prey, there is reason to believe that the prey can keep ahead in the race'. It is by no means clear what it might mean to say that one side or the other can 'keep ahead in the race'. It is probably true that, in any arms race, whenever either side produces a new adaptation it may enjoy something like a period of grace before the other side perfects a counter-adaptation. Individuals will experience this period of grace as a temporary decrease in the adverse effects of the other side. How long the period of grace lasts will depend on the various factors that govern rates of evolution. If one side consistently enjoys longer periods of grace than the other; if, say, the predator lineage is consistently slow to evolve a counter-adaptation to each relatively rapid prey adaptation, it might indeed be reasonable to talk of the prey 'keeping ahead in the race'.

Slobodkin's particular reasons for expecting prey to keep ahead in the race against predators can be applied the other way round, and we shall therefore not discuss them in detail. But his argument can be generalized in an interesting way.

It amounts to the suggestion that in an arms race between a specialist and a generalist the specialist will tend to win. The point is essentially the following economic one.

Adaptation budgets

Adaptations can be seen as 'costly', and resources that are used up in one adaptation are not available for use in another. A predator species that is adapted to hunt several different prey species is unlikely to be devastatingly effective against any one of them. As far as the individuals of any one prey species are concerned, this might be said to put them ahead in the arms race against this predator species. Similarly, if the prey invest maximally in adaptations for running fast, they may outrun predators which have to invest both in running speed and in adaptations for killing the prey once they have caught them. This is not to say it is in general 'better' to be a specialist than a generalist. In any particular arms race between a specialist and a generalist the specialist may have an advantage. But the generalist, by definition, runs many races and it may make a good living overall.

This kind of economic imbalance is also relevant when there are inequalities in strengths of selection pressures on the two sides. This is the subject of the next section.

The life–dinner principle (unequal selection pressures)

'The rabbit runs faster than the fox, because the rabbit is running for his life while the fox is only running for his dinner' (Dawkins 1979, after Aesop). There is a built-in imbalance between predator and prey with respect to the penalty of failure. Mutations that make foxes lose races against rabbits might therefore survive in the fox gene pool longer than mutations that cause rabbits to lose races can expect to survive in the rabbit gene pool. A fox may reproduce after losing a race against a rabbit. No rabbit has ever reproduced after losing a race against a fox. Foxes who often fail to catch prey eventually starve to death, but they may get some reproduction in first.

The life–dinner principle does not imply that 'rabbits' under strong selection pressure will win in the sense of driving weaker-selected 'foxes' to extinction. The suggestion is more subtle than this. As we saw in the last section, adaptations of one type, say for running fast, have to be paid for in the form of less perfect adaptations of other types. An illustration of the cost of antipredator adaptations is seen in some kinds of plants which produce chemical defence compounds. Within one species there tends to be a negative relation between investment in defence chemicals and fecundity (Price 1975). If 'foxes' are less strongly selected for running speed than 'rabbits', this means that they can afford to put more resources into other things such as gonads or intraspecific competitive adaptations. We could almost measure the cost of adaptations in 'gonad equivalents'. The life–dinner principle simply implies that, all other things being equal, animals under strong selection for speed will pay relatively more 'gonad equivalents' over to developing adaptations for running fast. Animals under weak selection for running speed will correspondingly benefit by being able to deploy relatively more resources in gonads or other adaptations.

Parker *et al.* (1972) used something like the life–dinner principle in discussing the origin of anisogamy. They postulated an ancient arms race between small, mobile 'male' gametes, and large, well nourished 'female' gametes. Selection favoured female gametes that discriminated against male ones, and simultaneously selection favoured male gametes that broke through the female defences. As Williams (1975) put it: 'This primeval conflict between the sexes was resolved in favor of the males because of more intense selection for male functions. Once the issue was decided, macrogametes abandoned the attempt to unite with other macrogametes and reject sperm'.

The general point here is that where one side in an arms race is under stronger selection pressure than the other, it will tend to have an advantage in that particular arms race. A similar idea was used by Pimentel (e.g. 1968) in his 'genetic feedback' theory of population regulation. The life–dinner principle is just one example of a way in which inequalities in rates of evolution may be important in determining the course and outcome of an arms race. A few others are mentioned in the next section.

General inequalities in potential rates of evolution

Palaeontologists agree that major taxonomic groupings vary in how fast they evolve (Simpson 1953). Stanley (1973) suggests that mammals evolve faster than bivalve molluscs because interspecific competition is fiercer for mammals. Schopf *et al.* (1975) feel that the difference may be more apparent than real, and Kauffman (1978) suggests that it is often not even apparent. But *if* it is true, for whatever reason, that mammals evolve on average nearly ten times as fast as bivalve molluscs in terms of observable phenotypes, would this give a mammal a built-in advantage in an arms race against a bivalve mollusc? The suggestion begs a serious question. It implies that bivalves are inherently incapable of evolving fast even when they are under strong selection pressure. But it is equally probable that lineages that evolve slowly do so precisely because they are *not* involved in severe arms races. Bivalves and frogs may have evolved slowly, not because they are incapable of evolving fast, but because they are getting on nicely as they are (Cain 1964).

On the other hand, there are some general reasons why some lineages might be inherently capable of responding to selection pressures more rapidly than others. These do not obviously differentiate mammals from bivalves or frogs, but they are worth brief consideration anyway. Population geneticists recognize various factors that in theory should influence rates of evolution, though any one of them might be ineffectual if other factors happened to be limiting. They include population size and genetic variance (Fisher 1930), tendency of the population to be broken up into semi-isolated sub-populations (Wright 1977), generation time, presence or absence of sexual reproduction, and perhaps mutation rate and recombination rate. For example, 'costly' though sex undoubtedly is, in the sense that a clone of asexual females is expected in the short term to outcompete sexual rivals within a species (Williams 1975), there can be little doubt that in an arms race a sexual lineage will have a good chance of out-evolving an asexual one (Maynard Smith 1978*a*).

Learning

It has often been pointed out that from a functional point of view there is a formal analogy between learning and natural selection (see, for example, Pringle 1951; Lorenz 1966). Both can be thought of as processes by which animals improve over time. In natural selection the improvement is seen only when a lineage of many individuals is viewed over many generations. In learning the improvement occurs during each individual's lifetime.

In an arms race between two non-learning species, one side may enjoy longer 'periods of grace' after producing each new adaptation, for any of the reasons that we have listed. But such periods of grace will be measured in tens or hundreds of generations. In an arms race where the individuals on one side are good at learning, there is a chance that they can cut the other side's period of grace down to less than one generation. On the other hand, the period of grace recurs for the same length of time in each generation, since unless there is cultural transmission every individual has to learn anew.

During the rest of this paper we shall refer back to these ways in which one side or the other in an arms race might have an advantage. We now turn to our two-way classification of arms races, beginning with an example of the most familiar class, interspecific asymmetric arms races.

Different types of arms race

Brood parasites and hosts: an asymmetric interspecific arms race

Interspecific asymmetric arms races include those between parasites and hosts, and predators and prey (Edmunds 1974). We shall discuss *brood parasitism* as an example, and shall use the arms race approach to discuss a particular paradox: why cuckoo hosts sometimes seem so clever, in discriminating against poorly mimetic eggs, and at other times seem so stupid in feeding young clearly different from their own (Zahavi 1979). Less anthropomorphically, how can a host provide the fastidious selection pressure that shapes the impressively accurate cuckoo egg mimicry, and a few days later be fooled by a cuckoo nestling which seems completely unlike its own young, and which may, indeed, be very much larger than the host itself?

The first thing to say is that the selection pressure in favour of discrimination by hosts is obviously stronger early in the incubation period than late. A host who realizes she has been duped as soon as the cuckoo egg is laid can destroy it immediately, or desert and begin a new nest having lost only a short time. When a baby cuckoo is old enough to impress humans as grotesquely different from the host's natural children, a host who at this point detects it gains little in comparison to one who continues to feed it; perhaps she gains only a few days, and it may be too late in the season to breed again anyway. She might even do better to finish the job and gain experience in the skills of parenthood. (We are tempted to go further and suggest that hosts who have little to lose might complete the rearing of cuckoos out of spite to other potential hosts, but unfortunately the idea seems theoretically unworkable.) Even if some of these suggestions are helpful, they only quantitatively mitigate our wonderment at the host's 'stupidity'. It still remains mysterious.

We prefer to think that the fundamental answer to the paradox lies in a subtle application of the arms race concept. When a human behaves in a way which is manifestly bad for him, for instance when he continually takes poison, we may explain his behaviour in at least two ways. He may not realize that the substance that he is drinking is poison, so closely does it resemble a genuinely nutritious substance. This corresponds to the host bird being fooled by the cuckoo's egg mimicry. Or he may be unable to save himself because of some direct subverting influence of the poison on his nervous system. Such is the case of the heroin addict who knows the drug is killing him, but who cannot stop taking it because the drug itself manipulates his nervous system. We have argued elsewhere that much of animal communication might profitably be regarded as akin to manipulation (Dawkins & Krebs 1978). There are indications that the orange gape and loud begging calls of the baby cuckoo have a supernormal effect (Wickler 1968). Could it be that the host can no more resist the supernormal manipulative power of the cuckoo nestling than the junkie can resist his fix, or than the brainwashed prisoner can resist the orders of his captor, however much it would benefit him to do so? Perhaps cuckoos have put their adaptive emphasis on mimetic deception at the egg stage, but on positive manipulation of the host's nervous system at the late nestling stage.

This is not a new idea. Wickler (1968) quotes Oskar Heinroth as having referred to the cuckoo as a 'vice of its foster parents' since they respond like 'addicts' to the supernormal stimulus. But the 'addiction' or 'manipulation' hypothesis has one outstanding difficulty which may be solved by an application of the arms race principle. It may be believable that an individual foster-parent could be hypnotized, addicted, or otherwise manipulated into feeding a cuckoo nestling which obviously bears no resemblance to one of her own. But surely, it will be said, over evolutionary time natural selection would put a stop to that: selection would favour hosts who have nervous systems resistant to the manipulation. Yes, but exactly the same thing could in theory be said of mimetic deception: mimicry cannot possibly evolve, because selection will always favour individuals who are not fooled by it! This, of course, is where the arms race idea comes in. Any nervous system can be subverted if treated in the right way. Any evolutionary adaptation of the host nervous system to resist manipulation by cuckoo nestlings simply lays itself open to counter-adaptation by the cuckoos. Selection acting on cuckoos will work to find whatever chinks there may be in the hosts' newly evolved psychological armour. Host birds may be very good at resisting psychological manipulation, but cuckoos might become even better at manipulating.

It does indeed seem easy to imagine that cuckoos might have a built-in advantage in their arms race with their hosts. The 'life–dinner principle' could apply: a cuckoo in the nest has *got* to manipulate its host successfully or it will surely die; its individual foster-parent will benefit somewhat if it resists manipulation, but it still has a good chance of future reproductive success in other years even if it fails to resist this particular cuckoo.

On the other hand, there is also evidence that under some conditions hosts may eventually get the upper hand. Systematic tests of the ability of various species of bird to discriminate against cuckoo eggs show that many non-parasitized species are excellent discriminators. Jourdain (1925) even suggested that non-parasitized species of European passerines are better discriminators than species used by cuckoos as hosts. While it is possible that cuckoos have simply avoided parasitizing species that are good discriminators, it is also possible that the good discriminators are species that used to be parasitized in the past. On this theory, cuckoos successfully parasitize a species for a while, then give it up when that host species becomes too good at discrimination. Maybe successive host species win their private arms race against cuckoo eggs, and the cuckoos then turn to new, 'evolutionarily naïve' victims. As C. E. Davies has pointed out to us in this connection, it may be significant that among the British gentes of cuckoo, which parasitize different species of host, egg mimicry ranges from nearly perfect in the meadow pipit gens to virtually non-existent in the hedge sparrow gens. Rothsteins' (1975*a*, *b*, 1976) evidence on the brown headed cowbird (*Molothrus ater*), an American brood parasite, can be interpreted in a similar way.

Our suggestion that cuckoos manipulate their hosts is not an alternative to the classical supernormal stimulus interpretation. Rather, we are providing a functional explanation of the host's maladaptive behaviour, instead of leaving it as an unexplained limitation of the nervous system.

The drug analogy is especially apt for insect parasites that use chemical means to coerce their hosts into acts that are profoundly damaging to their own fitness. The ant *Monomorium santschii* is a workerless parasite of *Monomorium salomonis*. An *M. santschii* queen enters a host nest in the same way as many other parasitic ants. But unlike parasitic queens of other species who directly kill the host queen, *M. santschii* achieves the same result by more subtle means. The host workers kill their own mother and adopt the usurper instead (Wilson 1971). It does not seem to be known why they do this, but we must infer that the parasite queen exerts a direct hold over their nervous systems. It might well be correct to label her weapon a pheromone, but it is probably more illuminating to think of it as a formidably powerful drug. In line with this interpretation, Wilson (1971) writes of symphylic substances as being 'more than just elementary nutritive substances or even analogues of the natural host pheromones. Several authors have spoken of a narcotizing effect of symphylic substances.' Wilson also uses the word 'intoxicant' and quotes a case in which worker ants under the influence of such a substance became temporarily disorientated and less sure of their footing.

Those who have never been brainwashed or addicted to a drug find it hard to understand their fellow men who are driven by such compulsions. In the same naïve way we cannot understand a host bird's being compelled to feed an absurdly oversized cuckoo, or worker ants wantonly murdering the only being in the whole world who is vital to their genetic success. But such subjective feelings are misleading, even where the relatively crude achievements of human pharmacology are concerned. With natural selection working on the problem, who would be so presumptuous as to guess what feats of mind control might not be achieved? Do not expect to see animals always behaving in such a way as to maximize their own inclusive fitness. Losers in an arms race may behave in some very odd ways indeed. If they appear disorientated and unsure of their footing, this may be only the beginning.

Intraspecific asymmetric arms races

The very idea of an arms race between members of the same species would have been considered quite revolutionary until a few years ago. Now we accept the idea even for parents and offspring, and members of a mated pair. In both cases R. L. Trivers (1972, 1974; Trivers & Hare 1976), building on the pioneering work of W. D. Hamilton (e.g. 1964, 1971) has been largely responsible for revolutionizing our way of thinking. We here concentrate on the arms race between parents and their children (Trivers 1974; Alexander 1974; Macnair & Parker 1978, 1979; Parker & Macnair 1978, 1979; O'Connor 1978).

Probably the most clearly worked out example of parent–offspring conflict is the case of sex ratios in social Hymenoptera (Trivers & Hare 1976; criticized by Alexander & Sherman (1977) and by Macnair (1978)). The example is particularly illuminating. The adaptation under discussion is that of the ratio of investment in male versus female reproductive offspring of the colony. We will consider only cases where there is a single queen who is the mother of all other individuals in the colony. If the queen has practical power over sex determination and parental investment, as would be expected if sociality evolved by parental domination (Alexander 1974; Michener & Brothers 1974), the evolutionarily stable investment ratio will be 1:1 as in the classical theory (Fisher 1930). But Trivers & Hare showed that, if the workers exert practical power over investment in young reproductives, the evolutionarily stable ratio will be 1:3. The difference results from the haplodiploid system of genetics which occurs in Hymenoptera (Hamilton 1964). Provided they are full sisters to the reproductives, workers are three times as closely related to female reproductives as to male ones. Since workers are sisters to the young reproductives in question, and the queen is the mother of the reproductives, the theory predicts a direct conflict between queens and workers.

This is a complex and subtle idea. It is misleading to think simply in terms of two rival ideal sex ratios, one for the queen and one for the workers; 1:1 and 1:3 are optimal ratios only in a particular sense. *If* the ratio of investment in the population as a whole is 1:1, selection will penalize queens who depart from it, and will favour workers who invest in only females rather than workers who invest 1:3. *If* the ratio of investment in the population as a whole is 1:3, selection will penalize workers who deviate from it, and will favour queens who lay only male eggs rather than queens who invest 1:1. From the present point of view the important fact is that there is no sex ratio that is stable if both queen and workers exert practical power over the ratio of investment in male and female reproductives.

Trivers & Hare concluded that, in most of the monogynous ant species that they studied, the workers seemed to have won the arms race and produced something close to the 1:3 ratio. The reason for victory, they suggested, is that the workers have more practical power: the workers actually look after the brood and are in a better position than the queen to control the amount of food given to male and female larvae. But there may be another more fundamental genetic reason related to the life–dinner principle, for worker 'victory'. This is that, when the sex ratio is 1:3 and workers are full siblings to reproductives, genes in workers are apparently being propagated more efficiently than genes in the queen would be at any conceivable sex ratio; conversely, when the sex ratio is 1:1, the 'optimum' from the queen's point of view, genes in the queen are only being propagated as efficiently as genes in workers. The queen's optimum, 1:1, is as good for the workers as it is for the queen. But the workers' own optimum, 1:3, is even better for them. Therefore genes for worker adaptations to outwit their mother have a built-in advantage over genes for queen adaptations to outwit her sterile daughters. Notice, incidentally, that these ideas go against the spirit of much of the language commonly used in discussions of social Hymenoptera, 'altruistic submission' by workers, for instance, and 'parental domination' by a 'super-beneficiary' queen.

Trivers & Hare's conclusions have been called into serious question by Alexander & Sherman (1977). The factual part of the argument will remain unresolved until more data have been obtained. But on the theoretical side there seems no doubt that Trivers & Hare's analysis is essentially correct (Charnov 1978), and not incompatible with Alexander & Sherman's alternative hypothesis.

We come now to the characteristically difficult aspect of intraspecific arms races. In an interspecific arms race like that between predator and prey, two entirely separate genetic lineages coevolve in parallel, mutally countering one another's adaptations, but in an intraspecific arms race the 'lineages' which are racing against each other are not really lineages at all. The genes that programme the development of queen behaviour are present in workers, and the genes that programme the development of worker behaviour are present in queens. All females are equipped with the same conditional rule or 'strategy' (Maynard Smith, this symposium): 'If fed like a queen invest 1:1 in male and female productives; if fed like a worker, invest 1:3', or whatever it is. Counter-adaptations and counter-counter-adaptations can still evolve in the arms race between queens and workers.

But it is an arms race between two branches of the same conditional strategy, not an arms race between two independent lineages with separate gene pools. We might see an organ in worker ants which is properly regarded as an adaptation for outwitting queens, and another organ in queens which is an adaptation for outwitting workers. Yet every female is genetically equipped to be either a queen or a worker.

This is a rather startling idea, which applies also to the arms race between males and females (Trivers 1972). As Parker (1979) has put it: '...selection can act in opposition on the two sexes. Commonly, for a given type of encounter, males will be favoured if they do mate and females if they don't'. We can therefore expect to find adaptations in males for overcoming female resistance, and counter-adaptations in females (Williams 1966). But unless the genes controlling these adaptations are carried on sex chromosomes, they will all have spent half their history in male bodies and half in female bodies.

To put the point in an exaggeratedly fanciful way, imagine that blueprints of Russian missiles spent half their time in Pentagon files, and blueprints of American counter-missiles spent half their time in Moscow! The analogy has now clearly gone too far, but a more realistic argument along these lines was developed by Alexander (1974). Briefly, he suggested that parents must always win the conflict against their offspring, because all children aspire to become parents. Any adaptation that helps a child to outwit his parents comes back and kicks him in the teeth when his own children use it against him. There seems no obvious reason why Alexander's argument should not be reversed (Dawkins 1976), and appropriate mathematical models (Macnair & Parker 1978; Parker & Macnair 1978) have failed to uphold it. The way to handle the problem is to postulate an asymmetric conditional strategy: 'if young do X, if old do Y', and see whether that conditional strategy is evolutionarily stable.

But we are here making the more general point that we need to think about whether, simply because only one genetic lineage is involved rather than two, intraspecific arms races will have quite different consequences from interspecific ones. One possible approach to answering the question would be along the lines of recent game theoretic analyses of aggressive behaviour for the special case where the combatants are genetic relatives (Treisman 1977; Grafen 1979; Maynard Smith 1978*b*).

Interspecific symmetric arms races

The subject of discussion here is competition between species, a major preoccupation of ecologists. We shall make no attempt to review this important subject, which is well covered in ecological text books. Here we are concerned simply to fit the topic into our two-way classification of arms races, and to point up some contrasts with other types of arms race.

When we look at a predator and its prey, or a parasite and its host, we seem to see numerous adaptations in each for foiling the adaptations of the other. Is this ever the case for interspecific competitors? Do we ever say of an organ or a behaviour pattern that it is an adaptation for outcompeting a rival species? By analogy with r and K selection, selection for mechanisms that prevent heterospecific competitors from gaining access to limited resources (called interference as opposed to exploitation competition) has been called α-selection (Emlen 1973). A well known ecological textbook published in 1978 states that 'The major evolutionary problem with α-selection is that the technique of interference will often affect members of the same species as well as members of competing species, so that competitive ability is achieved only by a reduction in the species' own values of r and K. An example would be a shrub which produces chemicals to retard the germination and growth of competing plants but which may suffer from autointoxication after several years.' By failing to distinguish individual autointoxication from intoxication of rival members of the same species, the author seems to fall victim to the implicit group-selectionism that lurks beneath the surface of much ecological writing. Nevertheless he is right that there are formidable problems in the way of interpreting any adaptation as a consequence of selection for interspecific competitive ability.

For example the suggestion by McKaye (1977, but see Coyne & Sohn 1978), that the cichlid fish *Cichlasoma nicaraguense* has been selected to rear predators of a severe competitor, could work only under very special circumstances. It is not enough that the presence of the predator in the lake benefits the species *C. nicaraguense* indiscriminately. Each foster-parent must benefit as an individual from the individual predators that it rears. This could happen if there is strong site attachment, as McKaye himself recognizes.

We may generalize this lesson. Unless there are special reasons to the contrary, we must expect that any individual who takes steps to reduce the competitive effectiveness of an interspecific rival is as likely thereby to help a conspecific rival as to benefit himself. To characterize selection for interspecific competitive adaptations as 'α-selection' is to miss the main point. The competition coefficient α is a measure of the impact of one species on the population size of another species. An adaptation that simply alters the population size of another species will not be favoured unless it benefits an *individual* possessing the adaptation as compared with another individual of the same species not possessing the adaptation. The coefficient α, like any other population parameter, conceals the individual-level interactions which really matter.

As we would expect from this argument, the most clearly recognizable adaptations for interspecific interference competition are those that directly benefit the competing individual: cannibalism by *Tribolium* beetles of interspecific rivals (Park 1962) benefits the individual cannibal not only by removing a competitor (this advantage also applies to the rest of the population) but also because the cannibal gains extra food. Apparent adaptations to interspecific competition are probably often extensions of intraspecific competitive adaptations. For example, Picman's (1977*a*, *b*) intriguing observation that nesting long-billed marsh wrens (*Telmatodytes palustris*) search out and peck eggs in nearby nests of other marsh wrens and of different species cannot be thought of as a purely interspecific adaptation. The same argument can be applied to interspecific territoriality (Cody 1974).

Even when adaptations to outcompete rival species do evolve, they will be unlikely to lead to prolonged and progressive arms races in the way predator–prey adaptations do. It is more probable that selection will favour evasion of competition, by niche separation. The theoretical question of how interspecific competition leads to niche separation and consequent reduction in competition has been analysed by Lawlor & Maynard Smith (1976) and Roughgarden (1976). They confirm the intuitive notion that individual level selection will favour species divergence. Direct evidence for evolutionary divergence as a result of competition comes from studies of 'character displacement' (Brown & Wilson 1956; Fenchel 1975; but see Grant 1972).

Intraspecific symmetric arms races

Intraspecific competition between individuals is the primary driving force of Darwinian evolution. Even adaptations to avoid being eaten by predators have evolved because individuals with such adaptations outcompeted conspecific individuals without them. This does not mean that all the arms races that we have so far discussed should truly be subsumed under the current heading. Improved crypticity in prey insects is truly the product of an interspecific arms race against bird predators in the following sense. If we could wave a wand and freeze evolution in the predators, further evolution of cryptic perfection in the prey would decelerate towards an optimum compromise with competing selection pressures and would eventually stop. It is because predators *evolve* that prey lineages go on evolving improved antipredator adaptations.

Nevertheless there are true intraspecific arms races, arms races that would continue even if with our magic wand we could freeze evolution in predators and all other species. An obvious example comes from male competition for females. Imagine a species in which large size is an advantage in male–male competition but not an advantage from any other point of view. It is entirely reasonable that competition will favour males that are slightly larger than the current population mode, *whatever the current mode may be* (Simpson 1953). This is a recipe for progressive evolution of the kind that we expect from an arms race. It is a true symmetric arms race, such as we shall probably never see in interspecific competition. Incidentally, female size may be 'dragged upwards' by selection for increased male size (Maynard Smith 1977).

An arms race of this kind could also be a recipe for eventual lineage extinction. It is plausible that many selection pressures will favour a fixed optimum, say a length of 10 cm, while sexual selection or other selection pressures favour a relative optimum: slightly larger than the current mode, whatever the current mode may be. If the 'relative' selection pressure causes the lineage to evolve sufficiently far away from the optimum size set by the 'absolute' selection pressures, the lineage will become increasingly vulnerable to extinction, until it is finally finished off by a comparatively minor environmental crisis.

This argument is not necessarily confined to the case of sexual selection. Hallam (1975) provides data relevant to this point for Jurassic bivalves and ammonites that exhibit Cope's Rule: the individuals get larger over evolutionary time. There is great variation in the rate at which they get larger. In the 60 palaeospecies surveyed by Hallam, the range was from 16 millidarwins to 3690 millidarwins (1 darwin was for practical purposes defined as a doubling in size per million years, but see Hayami 1978). Each of these lineages also had a *longevity*, measured from the time of its first appearance in the fossil record until its disappearance and presumed extinction. When the logarithm of the rate of evolution of a lineage, measured in millidarwins, was plotted against the logarithm of its longevity, measured in millions of years, a negative correlation was obtained. (It has been pointed out that Hallam's orginal way of plotting this relation may have exaggerated it since both axes are functions of one variable, lineage duration. But later analysis (Hallam 1978) shows that the negative correlation remains even after this possible artefact is removed.) It appears that the faster a lineage evolves large size, the sooner it goes extinct. This suggests, though of course it does not prove, that it is the large size itself that, in some sense, caused the extinction.

Hallam's own interpretation of his result is that selection, for whatever reason, in favour of large body size tends to lead to smaller population size and hence greater vulnerability to random extinction. Our suggestion, that the increased size is due to an intraspecific arms race in which selection favours individuals who are slightly larger than the current mode, and that the extinction is due to divergence of actual size from an absolute optimum, is compatible with Hallam's. Both may be true.

For these molluscs it is not obvious what the 'relative' selection pressure might have been. Ammonites are known to have been sexually dimorphic so sexual selection is conceivable for them. For bivalves, on the other hand, something analogous to Harper's (1977) suggestion for trees is more plausible. A tree that is competing for light will benefit by being slightly larger than its neighbours, however large they are. Similarly the filter-feeding current of a bivalve might be improved if the bivalve physically stands out above its neighbours.

By the way, it is an interesting suggestion that the explosive increase in brain size in our own relatively recent ancestors may have been the product of what amounts to an intraspecific arms race; the weapons being improved in this arms race were the skills of social manipulation and deception, countermanipulation and counterdeception (Trivers 1971; Humphrey 1976).

In concluding our two-way classification, we can summarize the most important general points as follows:

(*a*) Asymmetric interspecific arms races may result in what appears to be maladaptive behaviour, such as the feeding of cuckoos by hosts.

(*b*) Asymmetric intraspecific arms races may be fundamentally different from similar races between species, because the parties share a common gene pool.

(*c*) Symmetric interspecific arms races are likely to lead to evolutionary divergence rather than progressive escalation.

(*d*) Symmetric intraspecific arms races could lead to gradual progressive evolution and even extinction.

We now turn to a general question which applies to all arms races.

How do arms races end?

It is most unlikely that any single statement will suffice to describe how all arms races end. In this section we consider four general classes of endpoint, as follows. One side may drive the other to extinction; one side may reach a definable optimum, thereby preventing the other side from reaching its optimum; both sides may reach a mutual local optimum; or the race may persist in a theoretically endless limit cycle. Of these, the first two can be considered as ways in which one side 'wins'.

One side wins

The obvious way for an arms race to end in victory for one side is for the other side to be driven to extinction. This would be analogous to the laboratory observation by Pimentel *et al.* (1965) of a decline in housefly numbers brought about by an evolved increase in blowfly competitive ability. If a predator or parasite drove its victims extinct, it might as a result soon suffer a similar fate. Pathogens or parasites would at first sight seem likely candidates for this because, as Haldane (1949) said, '...microscopic and sub-microscopic parasites can evolve so much more rapidly than their hosts that the latter have little chance of evolving complete immunity to them'. Presumably he was thinking of their short generation time which, as suggested above, might give an advantage in any arms race. Most parasitologists believe, however, that parasites do not drive themselves to extinction through their own success. Instead they regard the long-term evolutionary trend as being a reduction rather than an escalation of the arms race: '...it is widely accepted that the longer an association (between parasite or pathogen and host) has been in existence in the evolutionary sense, the more likely it is to be a relationship unaccompanied by excesses on either side' (Sprent 1962). While we doubt the implicitly group-selectionist interpretations sometimes found in the parasitology literature, there may be other reasons for the parasite–host arms race not ending in extinction. For instance Pimentel's (1968) 'genetic feedback' or our own 'life–dinner principle' might be applicable here. Perhaps the best studied case history of an evolutionary arms race between pathogen and hosts is that between Australian rabbits and the introduced Myxoma virus. The evolution of a host–virus equilibrium has involved both a decline in viral virulence and an increase in host immunity (Burnet & White 1972).

The second way in which one side in an arms race can be said to 'win' is illustrated by Trivers & Hare's (1976) work on hymenopteran sex ratios, already mentioned. They concluded, in effect, that in many species, workers have won the arms race against queens. Whether or not they are right in this case, they have demonstrated in theory how it can be possible to assign victory to one side or another. They did it by making clear quantitative predictions that the evolutionarily stable investment ratio if the workers have practical power should be different from the ratio if the queen has practical power. Here the two 'optima' are incompatible. But as the next section shows, there may be cases of mutually compatible stable endpoints.

Equilibrium endpoints

The coevolution of plants and their pollinators is sometimes said to be an example of pure mutualism: plants give nectar 'in return' for having their pollen transferred to the stigma of another individual. But it is more realistic to think of a mutually exploitative arms race. Plants should be selected for meanness, to maximize the benefit derived from pollen transfer per unit of nectar cost (Heinrich 1975), while bees are designed to maximize their efficiency at harvesting nectar, measured as net nutritive gain per unit time (Pyke 1978, 1979*a*, *b*). Pyke shows us how both bees and plants might achieve a local optimum of efficiency, which could be a stable endpoint for the arms race.

Pyke studied the pollination of monkshood (*Aconitum columbianum*) and *Delphinium* by bumblebees, *Bombus appositus*. The flowers of both plants grow in a tall spike, the bottom flowers being older, larger, and female while the upper flowers on an inflorescence are smaller, younger, and male. The lower flowers tend to produce more nectar, so the bee gets the highest net reward per unit time at the bottom of each inflorescence. The bee's optimal harvesting strategy is probably to start at the bottom of each plant, and work its way upwards (without revisiting flowers) until the expected gain from staying in the same inflorescence drops to the gain expected by moving to the next plant (Charnov 1976; Pyke 1978). The bees do approximately this. At the same time the plants seem to have positioned their male and female flowers in such a way as to benefit from these bee movements. A bee arrives at the female flowers at the base of an inflorescence having just come from the male flowers at the top of the previous plant, so the chance of cross-fertilization is high.

Pyke goes further and suggests that the plants may practise a subtle form of exploitation of the bees' foraging behaviour. It might be thought that the optimal movement rule for a bee within an inflorescence would be to visit the flowers in the spiral sequence in which they are arranged, starting at the bottom and moving spirally upwards. However, Pyke's detailed observations of bee movements showed that instead the bee moves vertically up the inflorescence. It therefore misses some of the nectar-rich female flowers near the bottom of the spike, but Pyke was able to show that there is also a saving of time for the bee in flying directly upwards. This is because sideways moves cost more time than vertical moves. Calculations show that the bee cannot increase its net food intake per unit time by adopting any obvious rule other than moving straight upwards.

How might the plant benefit by the bees' vertical movement rule? Pyke suggests that bees normally do not carry enough pollen to fertilize every female flower on a plant. If the bee visited every single flower in the spiral sequence it would run out of pollen while still taking nectar from female flowers. This would be unfortunate for the plant. So it seems that the spiral arrangement of flowers might be no accident but a coevolved design feature maximizing the amount of outcrossing per unit of nectar expended. The coevolutionary race appears to be at a stable point in that a bee could not do better by adopting a new movement rule given the current structure of the plants, and a plant could not do better by adopting a different structure given the current behaviour of the bees.

A. Grafen has suggested to us that arms race theory may come to provide the dynamic counterpart to Maynard Smith's (1974) 'game' theory of evolutionarily stable states. But not all 'games' end in stable solutions.

Cyclic endings

Parker & Macnair (1979) and Parker (1979) have suggested that parent–offspring arms races and intersexual arms races may end neither in unilateral victory nor in stable equilibrium but in unresolvable cycling. In a genetic model of parent–offspring conflict, Parker & Macnair postulate a dominant 'conflictor' gene which causes children to demand more investment than the parental optimum (see also Macnair & Parker 1978, 1979; Parker & Macnair 1978). This is countered by the evolutionary spread of a 'suppressor' gene in parents, which causes them to apportion investment equally, regardless of the solicitations of individual young. The model assumes that 'conflictor' behaviour has a cost, and when most parents are 'suppressors' this cost is sufficient to reverse the former direction of selection on children: non-conflictor genes spread and the cycle resumes from its starting point. As Parker & Macnair emphasize, other models suggest that the parent–offspring arms race could, under other assumptions, end in evolutionary equilibrium. A similar example from pathogen–host arms races is given by Clarke (this symposium).

For the case of sexual conflict, Parker (1979) develops an endlessly cycling model which may be applicable to asymmetric arms races generally. We shall, indeed, illustrate his idea in a different context, that of an arms race between a prey and its predator. We follow Parker in thinking of it as a 'game', but our players are predator and prey rather than male and female. Each individual player pays a fixed, irretrievable cost. For example, the prey pay the cost of growing to a large size in order to be hard to swallow. And the predators pay the cost of growing to a large size too, the better to swallow them. There ensues an 'evolutionary chase' in which individuals on both sides pay successively higher costs as they get bigger and bigger. For each side there is an upper limit, determined by the value of 'winning', beyond which they will not further escalate the cost paid. The central idea is that this upper limit may be different for the two sides, for instance because of the 'life–dinner principle'. If the prey have a higher limit than the predators, what this would mean is that the prey are selected to invest further resources in getting bigger while analogous resources in the predators' case are diverted to other adaptations, say in growing more sexually attractive.

Parker's model suggests that the party that can afford to pay the higher cost will do so, whereupon selection will favour a rapid reduction in the cost paid by the other side. It will abruptly become smaller again, diverting more resources to other purposes. Theoretically this could lead to a return to the start of the cycle by both sides, though it remains to be seen whether anything like this happens in nature, and game theory suggests that in any case there may be a way out of the endlessly recycling chase.

As an aside, e.s.s. enthusiasts may like to note the essential difference between Parker's unstable game and the classical 'war of attrition' of Maynard Smith (1974). In the war of attrition both players pay the price 'bid' by the loser. In Parker's game, both pay the price they bid themselves. Growth to a particular size exacts irretrievable costs (see also Rose, in preparation).

Conclusion

Even more than when we began writing this paper, we are conscious of how superficial our understanding of evolutionary arms races is. What was planned as a short paper could, and probably should, be expanded into a book. We have tried to raise interesting questions even if we have not answered them. We come away with the feeling that the arms race concept could eventually provide the key to resolving some outstanding problems in evolutionary theory. We shall mention just three.

First, the idea that one side in an arms race might enjoy a built-in advantage over the other, say because of the 'life–dinner principle', could be used to account for many cases in which animals or plants appear to act against their own best interests. As in our example of the cuckoos manipulating their hosts, organisms on the losing side of an arms race may be manipulated into maximizing someone *else*'s inclusive fitness rather than their own (Dawkins 1978; Dawkins & Krebs 1978). The arms race idea provides one positive reason (there are, of course, many more; Maynard Smith 1978*b*; Gould & Lewontin, this symposium) for avoiding naïve perfectionism when we look at individual adaptation.

Secondly, the arms race concept may help to resolve the old controversy over whether lineages drive themselves to extinction through progressive evolution. The orthogenetic idea that evolution has its own inexorable internal momentum forcing lineages beyond the limits of natural selection is obviously absurd. But orthodox natural selection, *if there is an arms race*, can generate the kind of runaway process that looks like orthogenesis, and which might eventually hasten extinction of the lineage.

Thirdly, a related point, evolutionists repeatedly return to the question of whether there is any biological law which, in Maynard Smith's (1970) words, '...might enable us to put an arrow on time in evolutionary processes, as the second law of thermodynamics enables us to put an arrow on physical processes'. As Gould (1977) has argued in a stimulating historical essay, this is one of the 'eternal metaphors of paleontology'. He himself confesses to an anti-'directionalist', pro-'steady-statist' bias. He backs this up with evidence from the invertebrate fossil record although, as we have seen, he is prepared at times to offer directionalist interpretations of vertebrate history (Gould 1978). Maynard Smith, too, concludes on theoretical grounds that it is not easy to find a good basis for the commonsense feeling that there is temporal polarity in evolution, an arrow pointing in the direction of increasing complexity and even improved adaptedness.

Directionalist common sense surely wins on the very long time scale: once there was only blue-green slime and now there are sharp-eyed metazoa. But this is not the issue. Over the shorter term, say within the Cainozoic, is there a general trend which could in any sense be called improvement? Whatever attempts might be made to answer this question empirically, we suggest that even if there are no other theoretical reasons to expect such trends, arms races provide one.

If adaptation were solely to the inanimate environment, it is easy to believe that evolution would simply track Darwin's 'elements of air and water' in their random walk through time. Selection would be stabilizing until a change in the climate or an accidental geographical displacement introduced a brief interlude of directionality. Each such directional interlude would seem to be as likely to reverse as to continue the previous one. But in fact consistent directionality is introduced because the environment of any one evolving lineage includes other evolving lineages. Above all, it is because adaptations in one lineage call forth counter-adaptations in others, setting in motion the unstable evolutionary progressions we call arms races.

M. Ridley and P. Greenwood criticized a much longer first draft and advised on what to leave out. R. Anderson, S. Randolph and N. Knowlton gave additional advice. We benefited from various discussions at the conference on Evolutionarily Stable Strategies held in November 1978 at the Zentrum für interdisziplinäre Forschung of the University of Bielefeld.

References (Dawkins & Krebs)

Alexander, R. D. 1974 The evolution of social behavior. *A. Rev. Ecol. Syst.* **5**, 325–383.

Alexander, R. D. & Sherman, P. W. 1977 Local mate competition and parental investment in social insects. *Science, N.Y.* **196**, 494–500.

Brown, W. L. & Wilson, E. O. 1956 Character displacement. *Syst. Zool.* **5**, 49–64.

Burnet, F. M. & White, D. O. 1972 *The natural history of infectious diseases*, 4th edn. Cambridge University Press.

Cain, A. J. 1964 The perfection of animals. In *Viewpoints in biology*, vol. 3 (ed. J. D. Carthy & C. L. Duddington), pp. 36–63. London: Butterworth.
Charnov, E. L. 1976 Optimal foraging: the marginal value theorem. *Theor. Popul. Biol.* **9**, 129–136.
Charnov, E. L. 1978 Sex ratio selection in eusocial Hymenoptera. *Am. Nat.* **112**, 317–326.
Cody, M. L. 1974 *Competition and the structure of bird communities.* Princeton University Press.
Cott, H. B. 1940 *Adaptive coloration in animals.* London: Methuen.
Coyne, J. A. & Sohn, J. T. 1978 Interspecific brood care in fishes: reciprocal altruism or mistaken identity? *Am. Nat.* **112**, 447–450.
Darwin, C. R. 1859 *The origin of species.* London: John Murray.
Dawkins, R. 1976 *The selfish gene.* Oxford University Press.
Dawkins, R. 1978 Replicator selection and the extended phenotype. *Z. Tierpsychol.* **47**, 61–76.
Dawkins, R. 1979 The value judgements of evolution. In *Animal economics* (ed. M. A. H. Dempster & D. J. McFarland). London: Academic Press. (In the press.)
Dawkins, R. & Krebs, J. R. 1978 Animal signals: information or manipulation? In *Behavioural ecology* (ed. J. R. Krebs & N. B. Davies), pp. 282–309. Oxford: Blackwell Scientific Publications.
Edmunds, M. 1974 *Defence in animals.* New York: Longman.
Eldredge, N. & Gould, S. J. 1972 Punctuated equilibria: an alternative to phyletic gradualism. In *Models in paleobiology* (ed. T. J. M. Schopf), pp. 82–115. San Francisco: Freeman Cooper.
Emlen, J. M. 1973 *Ecology: an evolutionary approach.* Reading, Mass.: Addison Wesley.
Fenchel, T. 1975 Character displacement and coexistence in mud snails (Hydrobiidae). *Oecologia* **20**, 19–32.
Fisher, R. A. 1930 *The genetical theory of natural selection.* Oxford University Press.
Gould, S. J. 1977 Eternal metaphors of palaeontology. In *Patterns of evolution* (ed. A. Hallam), pp. 1–26. Amsterdam: Elsevier.
Gould, S. J. 1978 *Ever since Darwin.* London: Burnett-Deutsch.
Grafen, A. 1979 The hawk–dove game played between relatives. *Anim. Behav.* (In the press.)
Grant, P. R. 1972 Convergent and divergent character displacement. *Biol. J. Linn. Soc.* **4**, 39–68.
Haldane, J. B. S. 1949 Disease and evolution. In *Symposium on Ecological and Genetic Factors in Animal Speciation. Ricerca Scient. Suppl.* **19**, 68–75.
Hallam, A. 1975 Evolutionary size increase and longevity in Jurassic bivalves and ammonites. *Nature, Lond.* **258**, 493–496.
Hallam, A. 1978 How rare is phyletic gradualism and what is its evolutionary significance? *Paleobiology* **4**, 16–25.
Hamilton, W. D. 1964 The genetical evolution of social behaviour. I, II. *J. theor. Biol.* **7**, 1–52.
Hamilton, W. D. 1971 Selection of selfish and altruistic behavior in some extreme models. In *Man and beast: comparative social behavior* (ed. J. F. Eisenberg & W. S. Dillon), pp. 59–91. Washington: Smithsonian Institute.
Harper, J. L. 1977 *Population biology of plants.* London: Academic Press.
Hayami, I. 1978 Notes on the rates and patterns of size change in evolution. *Paleobiology* **4**, 252–260.
Heinrich, B. 1975 Energetics of pollination. *A. Rev. Ecol. Syst.* **6**, 139–170.
Humphrey, N. K. 1976 The social functions of intellect. In *Growing points in ethology* (ed. P. P. G. Bateson & R. A. Hinde), pp. **303–317.** Cambridge University Press.
Jerison, H. 1973 *Evolution of the brain and intelligence.* New York: Academic Press.
Jourdain, F. C. R. 1925 A study of parasitism in the cuckoos. *Proc. zool. Soc. Lond.*, pp. 639–667.

Kauffman, E. G. 1978 Evolutionary rates and patterns among Cretaceous Bivalvia. *Phil. Trans. R. Soc. Lond.* B **284**, 277–304.

Lawlor, L. R. & Maynard Smith, J. 1976 The coevolution and stability of competing species. *Am. Nat.* **110**, 79–99.

Lorenz, K. 1966 *Evolution and modification of behaviour*. London: Methuen.

McKaye, K. R. 1977 Defense of a predator's young by a herbivorous fish: an unusual strategy. *Am. Nat.* **111**, 301–315.

Macnair, M. R. 1978 An ESS for the sex ratio in animals, with particular reference to the social hymenoptera. *J. theor. Biol.* **70**, 449–459.

Macnair, M. R. & Parker, G. A. 1978 Models of parent–offspring conflict. I. Monogamy. *Anim. Behav.* **26**, 97–110.

Macnair, M. R. & Parker, G. A. 1979 Models of parent–offspring conflict. III. *Anim. Behav.* (In the press.)

Maynard Smith, J. 1970 Time in the evolutionary process. *Stud. Genet.* **23**, 266–272.

Maynard Smith, J. 1974 The theory of games and the evolution of animal conflicts. *J. theor. Biol.* **47**, 209–221.

Maynard Smith, J. 1977 Parental investment – a prospective analysis. *Anim. Behav.* **25**, 1–9.

Maynard Smith, J. 1978*a* *The evolution of sex*. Cambridge University Press.

Maynard Smith, J. 1978*b* Optimization theory in evolution. *A. Rev. Ecol. Syst.* **9**, 31–56.

Michener, C. D. & Brothers, D. J. 1974 Were workers of eusocial Hymenoptera initially altruistic or oppressed? *Proc. natn. Acad. Sci. U.S.A.* **71**, 671–674.

O'Connor, R. J. 1978 Brood reduction in birds: selection for fratricide, infanticide and suicide? *Anim. Behav.* **26**, 79–96.

Park, T. 1962 Beetles, competition, and populations. *Science, N.Y.* **138**, 1369–1375.

Parker, G. A. 1979 Sexual selection and sexual conflict. In *Reproductive competition and sexual selection* (ed. M. A. Blum & N. A. Blum), pp. 123–166. New York: Academic Press.

Parker, G. A., Baker, R. R. & Smith, V. G. F. 1972 The origin and evolution of gamete dimorphism and the male–female phenomenon. *J. theor. Biol.* **36**, 529–553.

Parker, G. A. & Macnair, M. R. 1978 Models of parent–offspring conflict. II. Promiscuity. *Anim. Behav.* **26**, 111–122.

Parker, G. A. & Macnair, M. R. 1979 Models of parent–offspring conflict. IV. Suppression: evolutionary retaliation by the parent. *Anim. Behav.* (In the press.)

Picman, J. 1977*a* Destruction of eggs by the long-billed marsh wren (*Telmatodytes palustris palustris*). *Can. J. Zool.* **55**, 1914–1920.

Picman, J. 1977*b* Intraspecific nest destruction in the long-billed marsh wren *Telmatodytes palustris palustris*. *Can. J. Zool.* **55**, 1997–2003.

Pimentel, D. 1968 Population regulation and genetic feedback. *Science, N.Y.* **159**, 1432–1437.

Pimentel, D., Feinberg, E. H., Wood, P. W. & Hayes, J. T. 1965 Selection, spatial distribution and the coexistence of competing fly species. *Am. Nat.* **99**, 97–109.

Price, P. W. 1975 *Insect ecology*. New York: Wiley.

Pringle, J. W. S. 1951 On the parallel between learning and evolution. *Behaviour* **3**, 174–215.

Pyke, G. H. 1978 Optimal foraging in bumblebees and coevolution with their plants. *Oecologia* **36**, 281–293.

Pyke, G. H. 1979*a* Optimal foraging in bumblebees: rule of movement between flowers within inflorescences. *Anim. Behav.* (In the press.)

Pyke, G. H. 1979*b* Optimal foraging in bumblebees: rule of departure from an inflorescence. In preparation.

Radinsky, L. 1978 Evolution of brain size in carnivores and ungulates. *Am. Nat.* **112**, 815–831.

Raup, D. M. 1977 Stochastic models in evolutionary palaeontology. In *Patterns of evolution* (ed. A. Hallam), pp. 60–78. Amsterdam: Elsevier.

Romanes, G. J. 1883 *Mental evolution in animals*. London: Kegan Paul Trench.

Rose, M. R. (in preparation) Evolutionary game arms races.

Rothstein, S. I. 1975*a* Evolutionary rates and host defenses against avian brood parasitism. *Am. Nat.* **109**, 161–176.

Rothstein, S. I. 1975*b* An experimental and teleonomic investigation of avian brood parasitism. *Condor* **77**, 250–271.

Rothstein, S. I. 1976 Experiments on defenses cedar waxwings use against cowbirds. *Auk* **93**, 675–691.

Roughgarden, J. 1976 Resource partitioning among competing species: a coevolutionary approach. *Theor. Popul. Biol.* **9**, 388–424.

Schopf, T. J. M., Raup, D. M., Gould, S. J. & Simberloff, D. S. 1975 Genomic versus morphologic rates of evolution: influence of morphologic complexity. *Paleobiology* **1**, 63–70.

Simpson, G. G. 1953 *The major features of evolution.* New York: Columbia University Press.

Slobodkin, L. B. 1974 Prudent predation does not require group selection. *Am. Nat.* **108**, 665–678.

Sprent, J. F. A. 1962 Parasitism, immunity and evolution. In *The evolution of living organisms* (ed. G. W. Leeper), pp. 149–165. Melbourne University Press.

Stanley, S. M. 1973 Effects of competition on rates of evolution with special reference to bivalve molluscs and mammals. *Syst. Zool.* **22**, 486–506.

Treisman, M. 1977 The evolutionary restriction of aggression within a species: a game theory analysis. *J. math. Psychol.* **16**, 167–203.

Trivers, R. L. 1971 The evolution of reciprocal altruism. *Q. Rev. Biol.* **46**, 35–57.

Trivers, R. L. 1972 Parental investment and sexual selection. In *Sexual selection and the descent of man* (ed. B. Campbell), pp. 136–179. Chicago: Aldine.

Trivers, R. L. 1974 Parent–offspring conflict. *Am. Zool.* **14**, 249–264.

Trivers, R. L. & Hare, H. 1976 Haplodiploidy and the evolution of the social insects. *Science, N.Y.* **191**, 246–263.

van Valen, L. 1973 A new evolutionary law. *Evol. Theory* **1**, 1–30.

Wickler, W. 1968 *Mimicry.* London: Weidenfeld & Nicholson.

Williams, G. C. 1966 *Adaptation and natural selection.* Princeton University Press.

Williams, G. C. 1975 *Sex and evolution.* Princeton University Press.

Wilson, E. O. 1971 *The insect societies.* Cambridge, Mass.: Harvard University Press.

Wright, S. 1977 *Evolution and the genetics of population.* Chicago University Press.

Zahavi, A. 1979 Parasitism and nest predation in parasitic cuckoos. *Am. Nat.* **113**, 157–159.

Proc. R. Soc. Lond. B **205**, 513–530 (1979)
Printed in Great Britain

The evolutionary genetics of sexual systems in flowering plants

By D. Charlesworth and B. Charlesworth
School of Biological Sciences, University of Sussex, Falmer, Brighton BN1 9QG, U.K.

Population genetic studies of the evolution of breeding systems in flowering plants are reviewed. The selective advantage of a gene's increasing the selfing rate is stressed. In the evolution of outbreeding mechanisms, some strong disadvantage to selfing must therefore be acting; it is suggested that this disadvantage is inbreeding depression. Populations with no absolute barrier to selfing, and with intermediate levels of self-fertilization, appear to be the most likely starting state for the evolution of outbreeding mechanisms. There is some evidence for inbreeding depression in such populations.

The evolution of distyly and dioecy are considered in some detail. An explanation for the existence of supergenes controlling these systems is proposed. The breakdown of distyly and tristyly are also considered. The evolution of recombination rates in selfing and outcrossing species is examined briefly.

Introduction

This paper reviews some applications of population genetic methods to the study of the evolution of plant breeding systems. It is often stated (e.g. Stebbins 1950, p. 155) that because the breeding system does not affect the individuals in which it is expressed, but only their descendents, any attempt to explain the evolution of breeding systems must examine their consequences in terms of possible advantages to the population. While it is true that some breeding system characteristics, for example variability or evolutionary flexibility (Mather 1943), are indeed meaningless in the context of the individual organism, there is no reason why we cannot treat the fundamental breeding system parameters, such as selfing rates or recombination frequencies, in conventional population genetic terms. We can study changes in frequencies of genes affecting these characters, and conditions for polymorphism, just as for any other characters subject to natural selection. This approach was used by Fisher (1930, 1941), but did not receive much attention from other workers until recently. Now, however, a number of problems in the evolution of breeding systems, many originally studied by Darwin (1876, 1877), have been subjected to study with the population genetics approach. Some of the methods used, and the results obtained, will be outlined below in relation to some specific problems in the evolution of sexual systems in the angiosperms. First, however, a number of general considerations will be introduced.

Selection on genes for self-fertilization

Probably the first explicit use of the population genetic method to solve a problem in breeding system evolution was Fisher's (1941) study of the spread of a gene increasing the selfing rate (defined as the fraction of ovules that are fertilized by pollen from the same plant). As will be seen, an understanding of this process illuminates several other problems relating to breeding systems. Fisher studied the case of a gene causing 100 % selfing (with intermediate dominance), introduced into an initially random-mating population of a hermaphrodite organism. He showed that (assuming no inbreeding depression) such a gene will always spread to fixation.

Nagylaki (1976) generalized this model to any degree of dominance, and also took into account the possibility that the selfing genotypes might use up pollen in selfing, thus reducing their ability to pollinate other individuals. He showed that the gene for selfing spreads unless the selfing individuals make no contribution to the outcrossing pollen pool. It is also easy to show that a mutation increasing the selfing rate from any starting value will increase to fixation, whatever its degree of dominance, assuming only that the decrease in contribution to the pollen pool is proportional to the increase in selfing rate with a constant of proportionality less than unity; a larger decrease seems unlikely.

Lloyd (1979*a*) studied a number of biologically realistic models of the way in which self-fertilization takes place, e.g. by selfing of those ovules that remain unfertilized after the flowers have been open for some time, and receptive to outcrossed pollen. Self-fertilization of this sort is unconditionally advantageous, whereas in Fisher's model (called 'prior selfing' by Lloyd, because only non-selfed ovules are available for outcrossing) a disadvantage to the progeny of selfing can prevent the spread of a gene for selfing. The disadvantage must be large, as selection for increased selfing is a strong force due to the fact that a selfing genotype can produce selfed progeny (with all its genes derived from the same parent) and can also compete with non-selfing plants for pollination of other individuals. For instance, assuming that all genotypes have equal contributions to the pollen pool for outcrossing, a dominant mutation giving complete selfing will increase in frequency by 50 % each generation, when introduced at low frequency into a random-mating population. It follows that, in considering the evolution of systems promoting outcrossing, it is necessary to seek for strong selective forces, capable of opposing this advantage of selfing.

Inbreeding depression and selfing rates in natural populations

Probably the major force opposing selection for increased selfing is inbreeding depression. The bad effects of inbreeding were first clearly documented by Darwin (1876). Evidence from natural populations is, however, very scanty. Although it is often claimed (e.g. Stebbins 1950, p. 165; Mather 1973, p. 131) that inbreeding species do not show inbreeding depression, we are aware of no study, other than of cultivated species, in which both selfing rates and inbreeding depression have been measured, although there is direct evidence for heterosis in *Arabidopsis thaliana* (Griffing & Langridge 1963), which has a natural rate of selfing of about 98 % (Snape & Lawrence 1971).

There appears to be indirect evidence for inbreeding depression in many species of plants that are capable of selfing under natural conditions. This evidence has accumulated as a by-product of studies designed primarily to estimate natural selfing rates by using marker genes, including allozyme loci detected by electrophoresis, to score the genotypes of the progeny of plants of known genotype. For a recent account of this technique, see Brown *et al.* (1978). Having estimated the selfing rate, it is simple to calculate the expected genotype frequencies, assuming that no selection is acting, and to compare them with those observed in the parental population. Assuming that the population is at equilibrium, and that the selfing rate has not changed since the parental population was formed, differences between the observed and expected genotype frequencies are evidence of selection.

In several such studies, heterozygotes were in excess of their expected frequencies (Marshall & Allard 1970; Harding 1970; Hamrick & Allard 1972; Brown *et al.* 1974; Harding & Barnes 1977; Phillips & Brown 1977; Ganders *et al.* 1977*a*; but see Babbel & Wain (1977) and Brown *et al.* (1978) for exceptions). Since heterozygote advantage appears to be uncommon (Lewontin 1974), it is unlikely that the heterozygote excess found in these studies is due to selection at the marker loci themselves, but is probably due to associations with chromosome segments having effects on viability, which can be generated in inbreeding populations (Haldane 1950; Bennett & Binet 1956; Ohta & Cockerham 1974). If this interpretation is correct (see Brown 1979), there is inbreeding depression for many chromosome regions in most partially selfing species studied. Unfortunately, these data do not permit estimates of its magnitude.

In theoretical work that includes inbreeding depression as a factor (see below), this has simply been treated as a fixed relative probability of survival of the products of selfing, compared with outcrossing. This ignores much real complexity, but is probably the best that can be done in view of our ignorance of the detailed causation of inbreeding depression, and of the difficulties in constructing realistic models (Cockerham & Rawlings 1967). A model with a parameter that changes over the generations was used by Maynard Smith (1977), but no attempt was made to ensure that the changes were quantitatively realistic.

The data also show that differences in selfing rates between populations are common, though it is not clear to what extent these differences are due to direct environmental effects, perhaps acting via the pollinator, and to what extent they represent adaptations by the plants to different environments (Allard *et al.* 1977). It seems probable that some genetic differences exist, as many predominantly selfing relatives of outcrossing species are known (see, for example, Vasek 1964, 1965, 1967; Ganders *et al.* 1977*a*, *b*; Rick *et al.* 1977).

Pollination

A second important factor to be considered in thinking about the evolution of breeding systems is the effectiveness of pollination. It is obvious that self-fertilization will have a further advantage, in addition to that discussed above, if individuals that tend not to self have difficulty in getting their ovules fertilized. It is therefore not unexpected that insect pollinated plants have often evolved devices to ensure selfing in areas where insects are scarce, for example in the Faroes (Hagerup 1951). In constructing population genetic models of outbreeding devices, one can combine inbreeding depression with this factor in a single parameter δ, defined as follows. Let $1-\delta$ be the probability that a selfed ovule results in a progeny individual that survives to the next generation, relative to the probability for a non-selfed ovule (which may or may not be fertilized). $1-\delta$ will be less than 1 if there is inbreeding depression causing loss of the progeny of selfed ovules, provided the chance of pollination for a non-selfed ovule is not too low.

Another situation that promotes self-fertilization is long distance dispersal. For example, populations on oceanic islands tend to be self-compatible (Baker 1955, 1967). This has been called 'Baker's rule', and is explicable in terms of a selective sieve that ensures that only self-compatible individuals can successfully establish a colony, unless the initial introduction consists of a number of individuals. A general review of these and other factors affecting the incidence of self-fertilization is given by Lloyd (1978).

Breakdown of heterostyly

This is a problem in which the major factors to be taken into account are the two that have been discussed above: the advantage of selfing, and the disadvantage of inbreeding depression. There are two types of heterostyly, distyly and tristyly, with two and three different forms of flowers, respectively, differing in their stigma and anther positions. Distyly occurs in at least 13 families, while tristyly is known in only 3 families. Darwin (1877) discovered that the flower forms are associated with a sporophytic incompatibility system which prevents fertilization of each form by other flowers of the same form (including self-fertilization). Both distyly and tristyly have broken down on numerous occasions. There are, for example, many self-compatible, monomorphic species in otherwise distylic genera such as *Primula*, *Linum* and *Limonium* (Darwin 1877; Baker 1959*b*, 1966), and also in tristylic genera such as *Lythrum* and *Oxalis* (East 1940; Ornduff 1972). The geographical distribution of these forms suggests that their appearance is connected with scarcity of insect pollinators, or with long-distance dispersal (Baker 1959*b*).

Distyly behaves as if the two flower forms were controlled by alleles at a single locus, with the long-styled form recessive in most cases (see Baker (1966) and Ornduff (1979) for exceptions), but the heterostyly 'loci' are in reality composites of at least three very tightly linked, but genetically separable, genes controlling the various characters by which the two forms differ. Although such 'supergenes' have been demonstrated genetically only in *Primula* (Ernst 1936*a*, *b*) and *Armeria* (Baker 1966), it is unlikely that a single locus controls all of the character differences between the forms; these are remarkably constant in different distylic species. Crossing over between the loci in the supergene can generate two complementary self-compatible 'homostyle' forms with stigma characters of one form and pollen type of the other. Other crossover types are possible, and have been detected as rare variants in some species (Ernst 1936*a*, *b*), but breakdown of distyly usually involves the fixation of one or other homostyle (Baker 1966).

Crosby (1949) derived equations for the changes in frequency of the three forms when 'long homostyles' are introduced into a population of the two normal distylic forms, neglecting crossing over between the three chromosome types. As expected from Fisher's result discussed above, homostyles always spread to fixation, unless the recombinant chromosome causing homostyly has a strongly deleterious effect (see also Bodmer 1960). Dowrick (1956) showed that short homostyles spread more slowly than long when each is introduced singly into a distylic population. It is possible to perform a similar analysis without the assumption of viability effects associated with any particular chromosome type, but merely that progeny produced by selfing have lower viability than those produced by outcrossing, i.e. assuming some inbreeding depression (B. & D. Charlesworth 1979). This seems a more plausible assumption. By using this model, it can be shown that self-fertile homostyles will not spread unless the δ parameter defined in the previous section is decreased below $\frac{1}{2}$, either because inbreeding depression has decreased, or because pollination of non-selfed ovules has become more uncertain. Long homostyles always have the highest selective advantage and outcompete all other breakdown forms if these are present initially at low frequency. This is because pollen from long homostyles is compatible with pin stigmas only; since pin is recessive the progeny can express the homostyle phenotype, whereas short homostyles produce pollen that is compatible with thrum, which is dominant, so fewer of the progeny express that phenotype. This could be the explanation for the predominance of breakdown to long homostyle (Ernst 1955; Ray & Chisaki 1957). The opposite should be true in the Plumbaginaceae, in which the dominance relations are reversed; this does, indeed, appear to be the case (Baker 1966).

The breakdown of tristyly occurs quite differently from that of distyly. Various different modes of breakdown have been found to occur (East 1940; Ornduff 1972). In particular, loss of the mid-styled form seems to happen more often than loss of the long or short forms (Ornduff 1972). Charlesworth (1979) studied a model of tristyly based on the two-locus system that is found in all three families that have tristyly (Fisher & Mather 1943; Weller 1976; S. H. C. Barrett, unpublished). Estimates of the strength of the 18 possible incompatibility reactions from several tristylic species, together with some assumptions about pollen transfer between the forms, suggests that mid may often be more selfing than the other forms. This leads to the result that if δ is less than $\frac{1}{2}$, the mid form tends to be present at high frequency, and increases with increased selfing, but the opposite is true when δ exceeds $\frac{1}{2}$; in this case, mid is frequently eliminated, leaving a dimorphic population with only long and short forms.

Evolution of heterostyly

The evolution of heterostyly is a much more complicated problem than its breakdown. In attempting to account for the evolution of distyly, we must seek a plausible sequence of steps that will lead to heterostyly from an initially self-compatible, homostyled state, and must account not only for the spread of the mutations involved, but also for their tight linkage. Since heterostyly is both a rare and complex system, it is perhaps to be expected that the pathway by which it evolved may have been complex (Fisher 1958, p. 41).

It is likely that morphological heterostyly evolved after the characteristic incompatibility system. Baker (1966) showed that species with a distylic type of incompatibility, but without distyly, are found in the Plumbaginaceae, and that more advanced species show the full complex of characters. It is also clear that at least two loci are required for the control of the pollen and stigma incompatibility reactions; this is proved by the fact that the distyly supergene can generate self-compatible crossover types with the pollen reaction of one form and the stigma reaction of the other. The first problem in accounting for the evolution of distyly is therefore to account for the incorporation of the two mutations required to give the incompatibility reaction. The first step could not have been a mutation in the stigma reaction, changing it from the ancestral condition to a new state, incompatible with the original form of pollen. Such a mutation would cause female sterility, and could not increase in frequency (D. & B. Charlesworth 1979*a*). The situation is quite different if we assume that the first change is a mutation to a new pollen type, so that self-fertilization cannot occur, and the ability to pollinate other individuals is also lost. Such a mutation will be at a considerable disadvantage, being effectively male-sterile, but can nevertheless spread if there is sufficiently high inbreeding depression, together with a high selfing rate (s), such that $s\delta$ exceeds $\frac{1}{2}$. This condition is likely to be satisfied only under exceptional circumstances.

Once a mutant type of pollen reaction has been established in a population, a mutation changing the stigma reaction, making it somewhat compatible with the new pollen type and incompatible with the original type, no longer causes female sterility, but can spread provided the mutant pollen type is common enough for the new female form to have a reasonable level of fertility. It is also necessary for this second mutation to occur at a locus that is linked to the gene controlling the pollen reaction. This can be seen from the following argument. With high δ, a new stigma type can have an advantage in individuals with the original pollen type, whose outcrossing rate will be increased, but will be disadvantageous in association with the new pollen type, as self-compatibility will then result. Now it is known from the general theory of two-locus systems that if a mutation occurs in a population polymorphic at another locus, and if it interacts favourably with one morph but unfavourably with the other, the spread of the mutation may depend on the recombination fraction between the loci (Bodmer & Parsons 1962). In the present case with a recessive pollen mutation segregating, a dominant mutation changing the stigma reaction will usually spread only if it occurs at a locus linked to that controlling the pollen reaction (D. &. B. Charlesworth 1979*a*). We have termed this type of effect a 'linkage constraint' (B. & D. Charlesworth 1978). The principle of a linkage constraint was probably known to Fisher (1931, p. 363). It may have been involved in other situations where supergenes have evolved (see below).

This explanation for the linkage between the incompatibility genes in distyly differs from the one that is usually given: that once genetic variation has been established at both loci, there will be selection for tighter linkage between them (Ford 1964, p. 91; Sheppard 1975, p. 87; Stebbins 1971, p. 10). In our model system, such selection could indeed be demonstrated, but genetic variation could usually not be established at both loci if they were initially unlinked.

Once an incompatibility system of the type described above has evolved, the problem arises of how (and why) morphological heterostyly evolves. The functional significance of heterostyly has usually been considered to lie in an increased chance of transfer of pollen to compatible individuals (Darwin 1877; Yeo 1975), or in an increased chance of a stigma receiving compatible pollen (Ganders 1974). But this cannot be the reason for the evolution of distyly, because if both types are initially homostyled, a new anther or stigma position does not increase the chance of pollen transfer to compatible stigmas, but would probably transfer less pollen to, or receive less from, other flowers. If the incompatibility reactions were imperfect, this disadvantage could be outweighed if a change in the anther position also reduced selfing. If the two forms differ in their level of self-incompatability, we might expect a mutation changing the anther or stigma position to have a net advantage in the more self-compatible form, but a disadvantage in the more self-incompatible one. Once again, therefore, a linkage constraint may operate. If the mutation can spread, it will become polymorphic in the population.

We may next ask which is more likely to be the initial morphological change in the evolution of heterostyly, anther or stigma position. It seems likely that a changed stigma position will usually be the first step. A mutant with a new stigma position may still receive sufficient pollen to ensure that any fertility loss is small, whereas a change in the anther position will lower the contribution to the pollen pool, and will therefore suffer much more from the disadvantage referred to above. This may explain why cases of distyly exist with only the stigma position varying (Darwin 1877; Dahlgren 1918, 1970; Baker 1958), whereas no case is known in which only the anther position varies.

The linkage constraint on the spread of a stigma-position mutation exists only if the mutation is dominant (D. & B. Charlesworth 1979*a*). If such a mutation arises, it can spread and be added to the linked pair of incompatibility genes. All the dominant alleles at the three loci become associated in one chromosome, and there is selection for tighter linkage between them. The final step required to produce complete heterostyly is a recessive mutation changing the anther position. There is no linkage constraint for such a mutation, but linked mutations increase faster than unlinked ones, and there is again selection for tighter linkage.

The sequence of steps outlined above for distyly to evolve is somewhat complex. The evolution of tristyly must be even more difficult to account for, though its occurrence in three families, with a similar genetic control in all three, suggests that the problem may not be insoluble. It is possible to construct a model for the evolution of tristyly with the observed type of two-gene control, by a sequence of three mutational steps, with fixation of one mutation (Charlesworth 1979). Each mutation must have pleiotropic effects on more than one character (e.g. stigma position and reaction). There is some evidence that the loci controlling tristyly may have such effects, though the apparent pleiotropy could also be explained if these loci are really supergenes (Ornduff 1972). There is, however, no genetic proof that this is so. If it were, the evolutionary problem would be even more difficult.

Evolution of gynodioecy and dioecy

Dioecy has probably evolved many times in flowering plants (Darwin 1877), most often from the hermaphrodite state, but also from monoecy (Lewis 1942; Lloyd 1972, 1975*a, b*) and from heterostyly (Baker 1959*a*; Opler *et al.* 1975; Lloyd 1979*b*). It is important to distinguish between these different origins, as the evolution of dioecy does not proceed in the same way from different starting states. We shall not say any more about the evolution of dioecy from heterostyly, which is not well understood, but will start with the simplest case, from hermaphroditism. In studying this problem, the factors mentioned previously (inbreeding depression and the advantage of selfing) are again important, but we must also take into account quantity of pollen and number of ovules produced. These are also important in the evolution of dioecy from monoecy, which we shall consider later.

Dioecy seems to evolve from the hermaphrodite state via an intermediate condition of polymorphism for females and hermaphodites (gynodioecy). This is consistent with the evident necessity for two genes, if a population is to go from the hermaphrodite state to one with males and females. It is, however, not obvious why the first step appears generally to be a mutation to complete male-sterility, rather than some more gradual process. Once gynodioecy is established, it may evolve into dioecy by a gradual (many genes with small effects) conversion of the hermaphrodites into males (Darwin 1877; Arroyo & Raven 1975). Thus the pollen-bearing sex is often highly variable in sex expression. We shall suggest reasons for these findings, and also for the fact that although two mutations (at least) are required, dioecy is often inherited as if by a single locus (see Westergaard 1958).

The first step is to study the conditions for the spread of male-sterility mutations. Lewis (1941) showed that for gene-controlled male-sterility to spread in a random-mating population, female fertility must be doubled, whereas any increase in female fertility ensures the spread of a cytoplasmic factor. It is possible that abolition of pollen production can give some automatic increase in ovule output (Darwin's (1877) 'reproductive compensation'), but a doubling seems unlikely. Lloyd (1974, 1975*b*, 1976) studied equilibria for male-sterility with various modes of inheritance, in populations with partial selfing of the hermaphrodites, and with inbreeding depression. If gynodioecy and dioecy are, as commonly thought, devices to ensure outcrossing, models in which the starting population has some degree of selfing are obviously required. Models with selfing have been studied by Ho & Ross (1974), Ross & Weir (1976), and B. & D. Charlesworth (1978). This assumption is consistent with the fact that dioecy and gynodioecy often occur in self-compatible groups, and that there is generally no self-incompatibility in the rare hermaphrodite individuals of dioecious or sub-dioecious species.

We have studied the conditions for spread of a dominant or recessive male-sterility mutation into a bisexual population with a rate of self-fertilization s (B. & D. Charlesworth 1978), by using a model similar to that of Lloyd. Such a mutation can spread if the advantages of avoiding inbreeding (due to the inability to self) and of any increased ovule output (due to compensation) outweigh the disadvantage of being unable to pollinate other plants. The condition for spread is that k, the increase in ovule output relative to a value of 1 for hermaphrodites, exceeds $1-2s\delta$. If this condition is satisfied, females will spread and reach an equilibrium frequency that cannot exceed $\frac{1}{2}$ (Lloyd 1974, 1975*b*, 1976).

It is simple to show that the analogous condition for the spread of a female-sterility mutation is that its increase in pollen output, relative to a value of 1 for hermaphrodites, must exceed $(1+s-2s\delta)/(1-s)$, which is evidently less easy to satisfy than the condition for male-sterility, especially with high selfing rates. Thus androdioecy (polymorphism for female sterility) is expected to be rare, as indeed it is (Darwin 1877). Lloyd (1975*b*) points out that the rarity of androdioecy provides support for models of dioecy in which the starting state is a population with some selfing. However, another factor that could make gynodioecy relatively commoner than androdioecy is the possibility for its establishment under cytoplasmic control; cytoplasmic inheritance of gynodioecy is very common (Lewis 1941, 1942). It is nevertheless clear that androdioecy is unlikely to be an intermediate stage in the evolution of dioecy. The only cases of androdioecy known to us are in andromonoecious species in which some purely staminate plants have been seen (in *Aesculus californicus* (H. G. Baker, personal communication), *Lomatium* (M. A. Schlessman, personal communication) and *Chionographis* (Hara 1968)). It is not known whether these are genetic polymorphisms.

Given an equilibrium population polymorphic for females and hermaphrodites, one can study the conditions for the spread of mutations reducing female fertility, i.e. converting the hermaphrodites into a more male form (B. & D. Charlesworth 1978). A necessary condition in that pollen output should be increased, but the increase required is less than when females are absent, and is lower the greater the ovule output of the females relative to that of hermaphrodites. The spread of genes making the male-fertile plants more male-like thus becomes increasingly easy, the more such mutations have already been incorporated. In addition, the spread of a dominant female-sterility factor into a population with recessive male-sterility depends on the linkage between the loci involved. There is a linkage constraint arising from the fact that a mutation causing reduced female fertility and increased pollen output can be advantageous only in the hermaphrodite plants. Unless the recombination fraction is less than some critical value, such a mutation cannot increase in frequency. This provides a possible explanation for the close linkage of the sex-controlling genes in dioecious plants (Westergaard 1958). If the first mutation is recessive and the second dominant, the males will be the heterozygous sex. If the opposite dominance relations hold, however, there is no linkage constraint and full dioecy cannot result unless tight linkage has been evolved after the establishment of the two-locus polymorphism. A few cases of female heterogamety are known, but it is not common (Westergaard 1958).

In studying the initial step in the evolution of dioecy, we assumed complete male-sterility, but we can use the same methods to consider mutations that reduce male fertility, but do not completely abolish it. Writing $1-K$ for the pollen output of such a mutant form, and s_f for its selfing rate, it can be shown that the condition for spread is that the female fertility $1+k$ must exceed

$$\frac{2(1-s\delta)-(1-K)(1-s)}{1+s_f(1-2\delta)}$$

(D. & B. Charlesworth 1978). If $\delta > \frac{1}{2}$, a given reduction in male fertility is advantageous if the selfing rate is reduced below some critical level determined by the values of the other parameters. This result suggests a reason for the fact that the first step in the evolution of dioecy from hermaphroditism is a mutation to complete male-sterility. In a hermaphrodite flower, with male and female organs in the same flower, a small reduction in pollen output may have little or no effect on the selfing rate, since the quantity of pollen produced in such flowers is usually greatly in excess of the amount that would be necessary to fertilize all of the ovules. Only a massive decrease in pollen production could reduce the selfing rate enough to satisfy the condition given above. Gynodioecy would therefore be unlikely to evolve gradually, but could arise in one step if an appropriate mutation occurred.

In a monoecious plant, however, a mutation changing the proportions of male and female flowers could be expected to alter the selfing rate by approximately the same proportion, and could therefore readily satisfy the condition for spread, expecially as such a change automatically increases the plant's overall female fertility. Even in the case of dioecy evolved from monoecy, however, we would predict that a small number of genes with large effects would have been involved, rather than many genes with extremely small effects, because the conditions for polymorphism for genes causing partial reductions in male fertility can be satisfied more easily, the larger the reduction in pollen output, K (D. & B. Charlesworth 1978). This is not surprising, since a complete male-sterility mutation cannot go to fixation, whereas one causing a smaller reduction clearly can do so.

Except for the difference in the nature of the initial reduction in male fertility when dioecy evolves from monoecy, subsequent events can proceed much as discussed for the hermaphrodite case. It can be shown that the presence of a 'female-like' form leads to increased selection for mutations reducing female fertility and increased male fertility in the original form, and that once again there is a linkage constraint on the spread of such mutations, if dominant. Dioecy could therefore evolve from monoecy by a sequence of mutations altering first female, then male fertility. Until full dioecy has evolved, which will require selection reducing recombination between the loci concerned, and is likely to take a long time, both sexes will show variation in sex expression, rather than just the 'males'. Lloyd (1973, 1979*c*) has termed this pathway 'paradioecy' and has pointed out its association with evolution from monoecy as the starting condition.

The model for the population genetics of genes with quantitative effects on male and female fertility can also be used to study the evolution of gynomonoecy, andromonoecy and monoecy (D. & B. Charlesworth 1978). It also provides a rigorous treatment of selection on pollen output with partial selfing, which leads to an explanation for the well documented fact of reduced pollen output in selfing species (Baker 1965; Cruden 1977).

Recombination rates

Another much studied breeding system parameter is the amount of recombination. Fisher (1930) was the first to propose models of individual, within-population selection for both decreased (pp. 116, 117) and increased (p. 118) recombination. Considerable attention has recently been given to the population genetics of modifiers of the amount of recombination that takes place between other genes. This field has been reviewed by Maynard Smith (1978).

In a random-mating population at equilibrium in a temporally and spatially constant environment, a recombination modifier introduced at low frequency will not spread unless the loci it affects interact epistatically in their effects on fitness in such a way that there is linkage disequilibrium. When there is such interaction, a modifier can spread only if it reduces recombination between the loci (Feldman 1972 and unpublished; B. & D. Charlesworth 1973). If there is some selfing, this is no longer true (Feldman & Balkau 1972). A number of models have been devised that generate selection for increased recombination in random-mating populations. They involve either 'hitch-hiking' effects produced by the spread of advantageous mutations (Fisher 1930, p. 118; Felsenstein & Yokoyama 1976; Strobeck *et al.* 1976) or non-uniformity of the environment in space or time (Williams 1975; Maynard Smith 1976*a*; Sturtevant & Mather 1938; Maynard Smith 1971; Charlesworth 1976).

It has been observed that chiasma frequencies tend to be increased in selfing species, compared with closely related outcrossing species (Stebbins 1950, pp. 180, 181, 1958; Garber 1956, 1958; Grant 1958; Lewis & John 1963; Zarchi *et al.* 1972; Arroyo 1973). Since the direct effect of inbreeding is probably to depress chiasma frequencies (Lamm 1963; Müntzing & Akdik 1948; Rees 1955; Dyal 1978), it appears likely that increased recombination has been selected for in such species. A number of population genetic models for selection on recombination have been shown to give increased recombination under selfing. All three classes of model mentioned above as giving selection for increased recombination can give increased selection when there is selfing (Charlesworth *et al.* 1977). With a constant environment, selfing can tend either to decrease (Feldman & Balkau 1972; Charlesworth *et al.* 1977) or to increase (Holden 1979; D. & B. Charlesworth 1979*b*) the recombination rate. The effect of selfing on recombination is complex, even in the two-locus cases studied so far, and may be even more so in multi-locus selective systems.

Heterozygosity for reciprocal translocation is a special type of device for reducing recombination, and is strongly correlated with high levels of selfing (Lewis & John 1963; James 1965). The classical example of *Oenothera*, in which reciprocal translations link up the entire genome in such a way that there is effectively no recombination (Cleland 1972), represents an extreme outcome of selection for reduced recombination with high selfing. It is tempting to explain this correlation in terms of models that lead to an increased selection pressure in favour of reduced recombination with high selfing rates; this can occur if, for example, there is heterozygote advantage at several loci and multiplicative fitness interactions between loci (Charlesworth *et al.* 1977). The occurrence of recessive or partially recessive deleterious mutations at many loci can also contribute to selection for translocation heterozygosity in selfing populations (de Waal Malefijt & Charlesworth 1979).

Discussion

We have tried to show how models for the evolution of breeding systems can be helpful in understanding and studying such systems, following Fisher's (1941) precept that 'unless individual advantage can be shown, natural selection affords no explanation of [characteristics] which appear to be beneficial to the species'. However, many problems remain. First, many details of those breeding systems whose evolution can be partly understood, are still unclear; for instance, the role of pleiotropy (and thus the number of genes involved) in the evolution of heterostyly. Secondly, the evolution of certain breeding systems still cannot be accounted for. There are difficulties in explaining monoecy and andromonoecy (D. & B. Charlesworth 1978), and the evolution of homomorphic incompatibility systems has scarcely been studied; there is not even a consensus about whether these are monophyletic or polyphyletic in origin (Whitehouse 1950; cf. Brewbaker 1959).

Finally, even when we can to some extent account for the evolution of particular systems, many broader patterns are difficult to account for. For example, there are correlations between chromosome number and plant type, and between life-form and breeding system (Stebbins 1950, 1971). It is not known for certain why dioecy is commoner in trees and shrubs than in herbs (Darwin 1876, p. 414; Baker 1959*a*; Bawa & Opler 1975), although an increased chance of self-pollination with a large number of flowers on the same plant is an attractive possibility, nor why dioecy and monoecy are associated with pollination by wind rather than insects (Darwin 1876, p. 422; but see Charnov *et al.* (1976) for a suggestion). Nor do we know why one outbreeding device rather than another evolved in any particular instance. For example, dioecy is commoner in New Zeland and Hawaii than in continental floras (Darwin 1876, p. 414; Carlquist 1966, 1974). Baker (1967) suggests that the reason may be that of various possible outcrossing devices, dioecy may evolve relatively easily (the mutations required are of types that occur often, and the conditions for their spread are not too hard to satisfy). If a flora consists mainly of self-compatible species, as is likely for the ancestors of the present-day floras of oceanic islands, then dioecy will probably be more likely to evolve than other outbreeding systems such as homomorphic or heteromorphic incompatibility. The problem of which type of outbreeding system will tend to evolve, has often been discussed in terms of which is most 'efficient' for the population; such a comparison has, for example, been made between dioecy and heterostyly (see, for example, Darwin 1877; Mather 1940). Such comparisons do not provide a satisfactory answer to this question, since they assume that selection between, rather than within, populations is the most important factor. It seems unlikely that this is often so (see Williams 1966; Maynard Smith 1976*b*).

References (Charlesworth & Charlesworth)

Allard, R. W., Kahler, A. L. & Clegg, M. T. 1977 Estimation of mating cycle components of selection in plants. In *Measuring selection in natural populations* (*Lecture notes in biomathematics*, vol. 19) (ed. F. B. Christiansen & T. M. Fenchel), pp. 1–19. Berlin: Springer-Verlag.

Arroyo, M. T. K. 1973 Chiasma frequency evidence on the evolution of autogamy in *Limnanthes floccosa* (Limnanthaceae). *Evolution* **27**, 679–688.

Arroyo, M. T. K. & Raven, P. H. 1975 The evolution of subdioecy in morphologically gynodioecious species of *Fuchsia* sect. *Encliandra* (Onagraceae). *Evolution* **29**, 500–511.

Babbel, G. R. & Wain, R. P. 1977 Genetic structure of *Hordeum jubatum*. I. Outcrossing rates and heterozygosity levels. *Can. J. Genet. Cytol.* **19**, 143–152.

Baker, H. G. 1955 Self-compatibility and establishment after 'long-distance' dispersal. *Evolution* **9**, 347–348.

Baker, H. G. 1958 Studies in the reproductive biology of West African Rubiaceae. *Jl W. Afr. Sci. Ass.* **4**, 9–24.

Baker, H. G. 1959*a* Reproductive methods as factors in speciation in flowering plants. *Cold Spring Harb. Symp. quant. Biol.* **24**, 177–190.

Baker, H. G. 1959*b* The contribution of autecological and genecological studies to our knowledge of the past migrations of plants. *Am. Nat.* **93**, 255–272.

Baker, H. G. 1965 Characteristics and mode of origin of weeds. In *Genetics of colonising species* (ed. H. G. Baker & G. L. Stebbins), pp. 147–168. New York: Academic Press.

Baker, H. G. 1966 The evolution, functioning and breakdown of heteromorphic incompatibility systems. I. The Plumbaginaceae. *Evolution* **20**, 349–368.

Baker, H. G. 1967 Support for Baker's Law – as a rule. *Evolution* **21**, 853–856.

Bawa, K. S. & Opler, P. A. 1975 Dioecism in tropical forest trees. *Evolution* **29**, 167–179.

Bennett, J. H. & Binet, F. E. 1956 Association between Mendelian factors with mixed selfing and random mating. *Heredity, Lond.* **10**, 51–55.

Bodmer, W. F. 1960 The genetics of homostyly in populations of *Primula vulgaris*. *Phil. Trans. R. Soc. Lond.* B **242**, 517–549.

Bodmer, W. F. & Parsons, P. A. 1962 Linkage and recombination in evolution. *Adv. Genet*, **11**, 1–100.

Brewbaker, J. L. 1959 Biology of the angiosperm pollen grain. *Indian J. Genet. Pl. Breed.* **19**, 121–133.

Brown, A. H. D. 1979 Enzyme polymorphism in plant populations. *Theor. Popul. Biol.* **15**, 1–42.

Brown, A. H. D., Marshall, D. R. & Albrecht, L. 1974 The maintenance of alcohol dehydrogenase polymorphism in *Bromus mollis* L. *Aust. J. biol. Sci.* **27**, 545–549.

Brown, A. H. D., Zohary, D. & Nevo, E. 1978 Outcrossing rates and heterozygosity in natural populations of *Hordeum spontaneum* Koch in Israel. *Heredity, Lond.* **41**, 49–62.

Carlquist, S. 1966 The biota of long distance dispersal. IV. Genetic systems in the floras of oceanic islands. *Evolution* **20**, 433–455.

Carlquist, S. 1974 *Island biology*. New York: Columbia University Press.

Charlesworth, B. 1976 Recombination modification in a fluctuating environment. *Genetics, Princeton* **83**, 181–195.

Charlesworth, B. & Charlesworth, D. 1973 Selection of new inversions in multi-locus genetic systems. *Genet. Res.* **21**, 167–183.

Charlesworth, B. & Charlesworth, D. 1978 A model for the evolution of dioecy and gynodioecy. *Am. Nat.* **112**, 975–997.

Charlesworth, B. & Charlesworth, D. 1979 The maintenance and breakdown of heterostyly. *Am. Nat.* (In the press.)

Charlesworth, D. 1979 The evolution and breakdown of tristyly. *Evolution*. **33**, 486–498.

Charlesworth, D. & Charlesworth, B. 1978 Population genetics of partial male-sterility and the evolution of monoecy and dioecy. *Heredity, Lond.* **41**, 137–153.

Charlesworth, D. & Charlesworth, B. 1979*a* A model for the evolution of heterostyly. *Am. Nat.* (In the press.)

Charlesworth, D. & Charlesworth, B. 1979*b* Selection for recombination in partially self-fertilizing populations. *Genetics, Princeton.* (In the press.)

Charlesworth, D., Charlesworth, B. & Strobeck, C. 1977 Effects of selfing on selection for recombination. *Genetics, Princeton* **86**, 213–226.

Charnov, E. L., Maynard Smith, J. & Bull, J. J. 1976 Why be a hermaphrodite? *Nature, Lond.* **263**, 125–126.

Cleland, R. 1972 *Oenothera. Cytogenetics and evolution.* New York: Academic Press.

Cockerham, C. C. & Rawlings, J. O. 1967 Apparent heterosis of a neutral gene with inbreeding. *Ciênc. Cul., S Paulo* **19**, 89–94.

Crosby, J. 1949 Selection of an unfavourable gene-complex. *Evolution* **3**, 212–230.

Cruden, R. W. 1977 Pollen–ovule ratios: a conservative indicator of breeding systems in flowering plants. *Evolution* **31**, 32–46.

Dahlgren, K. V. O. 1918 Heterostylie innerhalb der Gattung *Plumbago. Svensk bot. Tidskr.* **12**, 362–372.

Dahlgren, K. V. O. 1970 Heterostylie bei *Dyerophytum indicum.* (Gibbs ex Wright) O.K. (Plumbaginaceae). *Svensk bot. Tidskr.* **64**, 179–183.

Darwin, C. 1876 *Effects of cross and self fertilisation in the vegetable kingdom.* London: John Murray.

Darwin, C. 1877 *The different forms of flowers on plants of the same species.* London: John Murray.

Dowrick, V. P. J. 1956 Heterostyly and homostyly in *Primula obconica. Heredity, Lond.* **10**, 219–236.

Dyal, N. 1978 Inbreeding depression and heterosis in chromosome behaviour of radish (*Raphanus sativus* L. var *radicola* Pers.). *Curr. Sci.* **47**, 125–127.

East, E. M. 1940 The distribution of self-sterility in the flowering plants. *Proc. Am. phil. Soc.* **82**, 449–518.

Ernst, A. 1936*a* Heterostylie-forschung. Versuche zur genetischen Analyse eines Organisations und 'Anpassungs' merkmales. *Z. indukt. Abstamm.- u. VererbLehre* **71**, 156–203.

Ernst, A. 1936*b* Weitere Untersuchungen zur Phänanalyse zum Fertilitätsproblem und zur Genetik heterostyler Primeln. 2. *Primula hortensis* Wettstein. *Arch. Julius Klaus-Stift. VererbForsch.* **11**, 1–280.

Ernst, A. 1955 Self-fertility in monomorphic Primulas. *Genetica* **27**, 391–448.

Feldman, M. W. 1972 Selection for linkage modification. 1. Random mating populations. *Theor. Popul. Biol.* **3**, 324–346.

Feldman, M. W. & Balkau, B. 1972 Some results in the theory of three gene loci. In *Population dynamics* (ed. T. N. E. Greville), pp. 357–383. New York: Academic Press.

Felsenstein, J. & Yokoyama, S. 1976 The evolutionary advantage of recombination. II. Individual selection for recombination. *Genetics, Princeton* **83**, 845–859.

Fisher, R. A. 1930 *The genetical theory of natural selection.* Oxford: Clarendon Press.

Fisher, R. A. 1931 The evolution of dominance. *Biol. Rev.* **6**, 345–368.

Fisher, R. A. 1941 Average excess and average effect of a gene substitution. *Ann. Eugen.* **11**, 53–63.

Fisher, R. A. 1958 *The genetical theory of natural selection*, 2nd edn. New York: Dover.

Fisher, R. A. & Mather, K. 1943 The inheritance of style lengths in *Lythrum salicaria. Ann Eugen.* **11**, 31–38.

Ford, E. B. 1964 *Ecological genetics.* London: Methuen.

Ganders, F. R. 1974 Disassortative pollination in the distylous plant *Jepsonia heterandra. Can. J. Bot.* **52**, 2401–2406.

Ganders, F. R., Carey, K. & Griffiths, A. J. F. 1977*a* Natural selection for a fruit dimorphism in *Plectritis congesta* (Valerianaceae). *Evolution* **31**, 873–881.

Ganders, F. R., Carey, K. & Griffiths, A. J. F. 1977*b* Outcrossing rates in natural populations of *Plectritis brachystemon* (Valarianaceae). *Can. J. Bot.* **55**, 2070–2074.

Garber, E. D. 1956 The genus *Collinsia.* I. Chromosome number and chiasma frequency of species in the two sections. *Bot. Gaz.* **118**, 71–73.

Garber, E. D. 1958 The genus *Collinsia.* VII. Additional chromosome numbers and chiasma frequencies. *Bot. Gaz.* **120**, 55–56.

Grant, V. 1958 The regulation of recombination in plants. *Cold Spring Harb. Symp. quant. Biol.* **23**, 337–363.

Griffing, B. & Langridge, J. 1963 Phenotypic stability of growth in the self-fertilized species, *Arabidopsis thaliana*. In *Statistical genetics and plant breeding* (ed. W. F. Hanson & H. F. Robinson), pp. 363–394. Washington, D.C.: National Academy of Science – National Research Council.

Hagerup, O. 1951 Pollination in the Faroes – in spite of rain and poverty of insects. *Biol. Meddr.* **18**, 3–48.

Haldane, J. B. S. 1950 The association of characters as a result of inbreeding and linkage. *Ann. Eugen.* **15**, 15–23.

Hamrick, J. L. & Allard, R. W. 1972 Microgeographical variation in allozyme frequencies in *Avena barbata*. *Proc. natn. Acad. Sci. U.S.A.* **69**, 2100–2104.

Hara, H. 1968 A revision of the genus *Chionographis*. *J. Jap. Bot.* **43**, 257–267.

Harding, J. 1970 Genetics of *Lupinus*. II. The selective disadvantage of the pink flower color mutant in *Lupinus nanus*. *Evolution* **24**, 120–127.

Harding, J. & Barnes, K. 1977 Genetics of *Lupinus*. X. Genetic variability, heterozygosity and outcrossing in colonial populations of *Lupinus succulentus*. *Evolution* **31**, 247–255.

Ho, T-Y. & Ross, M. D. 1974 Maintenance of males and females in hermaphrodite populations. *Heredity, Lond.* **32**, 113–118.

Holden, L. R. 1979 New properties of the two-locus partial selfing model with selection. *Genetics*. (In the press.)

James, S. H. 1965 Complex hybridity in *Isotoma petraea*. I. The occurrence of interchange heterozygosity, autogamy and a balanced lethal system. *Heredity, Lond.* **20**, 341–353.

Lamm, R. 1936 Cytological studies in inbred rye. *Hereditas* **22**, 217–240.

Lewis, D. 1941 Male sterility in natural populations of hermaphrodite plants. *New Phytol.* **40**, 56–63.

Lewis, D. 1942 The evolution of sex in flowering plants. *Biol. Rev.* **17**, 46–67.

Lewis, K. R. & John, B. 1963 *Chromosome marker*. London: Churchill.

Lewontin, R. C. 1974 *The genetic basis of evolutionary change*. New York: Columbia University Press.

Lloyd, D. G. 1972 Breeding systems in *Cotula*. II. Monoecious populations. *New Phytol.* **71** 1195–1202.

Lloyd, D. G. 1973 Sexual dimorphism in *Cotula* (Compositae). (Abstr.) *Genetics, Princeton* **74**, s161.

Lloyd, D. G. 1974 Theoretical sex ratios of dioecious and gynodioecious angiosperms. *Heredity, Lond.* **32**, 11–34.

Lloyd, D. G. 1975*a* Breeding systems in *Cotula*. III. Dioecious populations. *New Phytol.* **74**, 109–123.

Lloyd, D. G. 1975*b* The maintenance of gynodioecy and androdioecy in angiosperms. *Genetica* **45**, 325–339.

Lloyd, D. G. 1976 The transmission of genes via pollen and ovules in gynodioecious angiosperms. *Theor. Popul. Biol.* **9**, 299–316.

Lloyd, D. G. 1978 Demographic factors and self-fertilization in plants. In *Demography and the dynamics of plant populations* (ed. O. T. Solbrig), pp. 67–88. Oxford: Blackwell.

Lloyd, D. G. 1979*a* Some reproductive factors affecting the selection of self-fertilization in plants. *Am. Nat.* **113**, 67–79.

Lloyd, D. G. 1979*b* Evolution towards dioecy in heterostylous populations. *Pl. Syst. Evol.* **131**, 71–80.

Lloyd, D. G. 1979*c* The distribution of gender in four representative angiosperms. In preparation.

Marshall, D. R. & Allard, R. W. 1970 Maintenance of isozyme polmorphism in natural populations of *Avena barbata*. *Genetics* **66**, 393–399.

Mather, K. 1940 Outbreeding and separation of the sexes. *Nature, Lond.* **145**, 484–486.

Mather, K. 1943 Polygenic inheritance and natural selection. *Biol. Rev.* **18**, 32–64.

Mather, K. 1973 *The genetical structure of populations*. London: Chapman & Hall.

Maynard Smith, J. 1971 What use is sex? *J. theor. Biol.* **30**, 319–395.

Maynard Smith, J. 1976*a* A short-term advantage for sex and recombination through sib-competition. *J. theor. Biol.* **63**, 245–258.

Maynard Smith, J. 1976*b* Group selection. *Q. Rev. Biol.* **51**, 277–283.

Maynard Smith, J. 1977 The sex habit in plants and animals. In *Measuring selection in natural populations* (*Lecture notes in biomathematics*, vol. 19) (ed. F. B. Christiansen & T. M. Fenchel), pp. 315–331. Berlin: Springer-Verlag.

Maynard Smith, J. 1978 *The evolution of sex*. Cambridge University Press.

Müntzing, A. & Akdik, S. 1948 Cytological disturbances in the first inbred generations of rye. *Hereditas* **34**, 485–509.

Nagylaki, T. 1976 A model for the evolution of self-fertilization and vegetative reproduction. *J. theor. Biol.* **58**, 55–58.

Ohta, T. & Cockerham, C. C. 1974 Detrimental genes with partial selfing and effects on a neutral locus. *Genet. Res.* **23**, 191–200.

Opler, P. A., Baker, H. G. & Frankie, G. W. 1975 Reproductive biology of some Costa Rica *Cordia* species (Boraginaceae). *Biotropica* **7**, 234–247.

Ornduff, R. 1972 The breakdown of trimorphic incompatibility in *Oxalis* section *Corniculatae*. *Evolution* **26**, 52–65.

Ornduff, R. 1979 The genetics of heterostyly in *Hypericum aegypticum*. *Heredity, Lond.* **42**, 271–272.

Phillips, M. A. & Brown, A. H. D. 1977 Mating system and hybridity in *Eucalyptus pauciflora*. *Aust. J. biol. Sci.* **30**, 337–344.

Ray, P. M. & Chisaki, H. F. 1957 Studies on *Amsinckia*. II. Relationships among the primitive species. *Am. J. Bot.* **44**, 537–544.

Rees, H. 1955 Genotypic control of chromosome behaviour in rye. 1. Inbred lines. *Heredity, Lond.* **9**, 93–116.

Rick, C. M., Fobes, J. F. & Holle, M. 1977 Genetic variation in *Lycopersicon pimpinellifolium*: evidence of evolutionary change in mating systems. *Pl. Syst. Evol.* **127**, 139–170.

Ross, M. D. & Weir, B. S. 1976 Maintenance of males and females in hermaphrodite populations and the evolution of dioecy. *Evolution* **30**, 425–441.

Sheppard, P. M. 1975 *Natural selection and heredity*. London: Hutchinson.

Snape, J. W. & Lawrence, M. J. 1971 The breeding systems of *Arabidopsis thaliana*. *Heredity, Lond.* **27**, 299–302.

Stebbins, G. L. 1950 *Variation and evolution in plants*. New York: Columbia University Press.

Stebbins, G. L. 1958 Longevity, habitat and release of genetic variability in the higher plants. *Cold Spring Harb. Symp. quant. Biol.* **23**, 365–377.

Stebbins, G. L. 1971 *Chromosomal evolution in higher plants*. London: Edward Arnold.

Strobeck, C., Maynard Smith, J. & Charlesworth, B. 1976 The effects of hitch-hiking on a gene for recombination. *Genetics, Princeton* **82**, 547–558.

Sturtevant, A. H. & Mather, K. 1938 The interrelations of inversions, heterosis and recombination. *Am. Nat.* **72**, 447–452.

Vasek, F. C. 1964 Outcrossing in natural populations. I. The Breckinridge Mountain population of *Clarkia exilis*. *Evolution* **18**, 213–218.

Vasek, F. C. 1965 Outcrossing in natural populations. II. *Clarkia unguicultata*. *Evolution* **19**, 152–156.

Vasek, F. C. 1967 Outcrossing in natural populations. III. The Deer Creek population of *Clarkia exilis*. *Evolution* **21**, 241–248.

de Waal Malefijt, M. & Charlesworth, B. 1979 A model for the evolution of translocation heterozygosity. *Heredity, Lond.* (In the press.)

Weller, S. G. 1976 The genetic control of tristyly in *Oxalis* section *Ionoxalis*. *Heredity, Lond.* **37**, 387–393.

Westergaard, M. 1958 The mechanism of sex determination in dioecious flowering plants. *Adv. Genet.* **9**, 217–281.

Whitehouse, H. L. K. 1950 Multiple allelomorph incompatibility of pollen and style in the evolution of angiosperms. *Ann. Bot.* **14**, 199–216.

Williams, G. C. 1966 *Adaptation and natural selection: a critique of some current evolutionary thought*. Princeton University Press.
Williams, G. C. 1975 *Sex and evolution*. Princeton University Press.
Yeo, P. F. 1975 Some aspects of heterostyly. *New Phytol.* **75**, 147–153.
Zarchi, Y., Simchen, G., Hillel, J. & Schaap, T. 1972 Chiasmata and the breeding system in wild populations of diploid wheats. *Chromosoma* **38**, 77–94.

Proc. R. Soc. Lond. B **205**, 531–546 (1979) **531**
Printed in Great Britain

The evolution of ageing and longevity

By T. B. L. Kirkwood† and R. Holliday, F.R.S.‡
† *National Institute for Biological Standards and Control, Hampstead, London NW3 6RB, U.K.*
‡ *National Institute for Medical Research, Mill Hill, London NW7 1AA, U.K.*

Ageing is not adaptive since it reduces reproductive potential, and the argument that it evolved to provide offspring with living space is hard to sustain for most species. An alternative theory is based on the recognition that the force of natural selection declines with age, since in most environments individuals die from predation, disease or starvation. Ageing could therefore be the combined result of late-expressed deleterious genes which are beyond the reach of effective negative selection. However, this argument is circular, since the concept of 'late expression' itself implies the prior existence of adult age-related physiological processes. Organisms that do not age are essentially in a steady state in which chronologically young and old individuals are physiologically the same. In this situation the synthesis of macromolecules must be sufficiently accurate to prevent error feedback and the development of lethal 'error catastrophes'. This involves the expenditure of energy, which is required for both kinetic proof-reading and other accuracy promoting devices. It may be selectively advantageous for higher organisms to adopt an energy saving strategy of reduced accuracy in somatic cells to accelerate development and reproduction, but the consequence will be eventual deterioration and death. This 'disposable soma' theory of the evolution of ageing also proposes that a high level of accuracy is maintained in immortal germ line cells, or alternatively, that any defective germ cells are eliminated. The evolution of an increase in longevity in mammals may be due to a concomitant reduction in the rates of growth and reproduction and an increase in the accuracy of synthesis of macromolecules. The theory can be tested by measuring accuracy in germ line and somatic cells and also by comparing somatic cells from mammals with different longevities.

Introduction

The process of ageing is not intrinsic to all living organisms. Bacteria and many eukaryotic microorganisms can divide indefinitely; many higher plants are capable of unlimited vegetative propagation, and some simple animals, such as coelenterates and *Turbellaria*, have powers of regeneration and renewal which apparently allow them to escape senescent changes (see Comfort 1964). Moreover, the germ line of all sexually reproducing organisms is potentially immortal.

The ageing of individual organisms is most obvious in the more highly evolved animals. It can be defined as a rapid increase in the force of mortality after a period of time which is characteristic for each species. The survival curve of a population of animals that age is shown in figure 1, curve *d*. On the other hand, if the probability of death is constant with time, then survival is exponential, as shown in figure 1, curve *a*. Since the rate of ageing varies between species or can be strongly

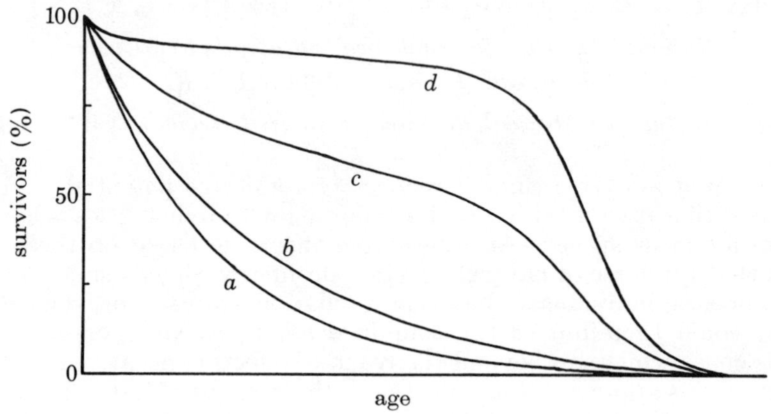

FIGURE 1. Contrasting survival curves for four hypothetical populations. Curve *a* shows exponential survival with constant mortality at all ages. Curve *b* closely resembles the survival of many wild populations. Curve *c* represents the survival of slowly reproducing mammals in a natural environment. Curve *d*, exhibiting a sharp increase in mortality with age, describes protected populations, such as laboratory animals or humans.

affected by the genetic background within a single species, we know that it is genetically determined; therefore, the process of ageing itself must be subject to natural selection. It would be a mistake to believe that somatic ageing is inevitably associated with the complexity of a highly evolved organism. As Williams (1957) has pointed out, the problems of evolving the means to develop such organisms from a single cell must be enormously greater than the mere maintenance of the adult organism by the repair or renewal of cells or tissues.

Non-adaptive nature of ageing

The problem is to understand why ageing came to evolve at all, as it is clear that it is not in itself adaptive. An animal that grows to maturity and reproduces indefinitely thereafter has a greater fitness than one that grows to maturity and then reproduces for only a fixed period of time. The first attempt to explain evolution of ageing was made by Weismann (1891). If evolution is to occur at all, then generation must follow generation and he proposed that the process of ageing was beneficial, since it removed old and worn-out individuals from the population, thus providing their progeny with living space and allowing natural selection to take place. There are three strong arguments against his view. The theory is circular in that it presupposes what it sets out to explain, namely that individuals wear out as they get older. Moreover, in most natural populations old and worn-out individuals are not in any case frequently seen, since animals normally die from predation, disease or starvation (see, for example, Lack 1954). Their survival in the wild is much closer to an exponential curve (figure 1, curve *b*) than the survival curve seen for animals kept in the laboratory under very favourable environmental conditions (figure 1, curve *d*). Finally, Weismann's argument depends on group selection and therefore suffers from all the difficulties associated with this theory. For example, in a population of animals that age, an individual with a mutation increasing longevity will have a selective advantage. Therefore, the process of ageing should not be stably maintained in the population. Nevertheless, there may be special cases where Weismann's general argument may be valid and we shall return to these later.

Ageing as a consequence of declining force of selection

In recognition of the failure of Weismann's adaptive argument to provide a satisfactory explanation of ageing, Medawar (1952) proposed a completely different theory based on the hypothesis that harmful genes whose time of expression lay beyond the greater part of the natural range of lifespan could accumulate with little or no check. The basis of Medawar's hypothesis was the crucial realization that, *even in the absence of ageing*, the effect of mortality from other causes would be to reduce the force of natural selection in each successive portion of the lifespan. The reason for this is that the proportion of newborn individuals expected to survive and reproduce past age x must always be less than the proportion surviving and reproducing past age $x-1$. As a result, natural selection would generally favour the advancement of beneficial genes into an earlier stage of the lifespan where they would be of greater advantage, and would defer the expression of deleterious genes to the point where they would hardly ever have the opportunity to do harm. He concluded that ageing is the result of the cumulative phenotypic expression of deleterious genes in one individual that lives much longer than the average lifespan of the species in its natural environment.

Williams (1957) extended Medawar's concept to include the possibility of pleiotropic genes having the potential for both good and bad effects at different times. By a similar argument, he suggested that natural selection would have brought forward the good effects and delayed the bad ones, and he proposed that ageing was an amalgam of the late bad effects of pleiotropic genes that were retained on the strength of their early adaptive benefits. In a sense, Williams's idea was counter to his observation, cited above, that in comparison with the feat of morphogenesis it should be relatively simple to maintain what is already formed. For example, it should be a simple matter to suppress the expression of the pleiotropic genes after their good effects have been generated and before their bad effects become manifest. The reason why this need not always happen is that the bad effects may occur too late to be selected against.

The precise manner in which the force of natural selection declines with increasing age is determined by combination of the patterns of mortality and reproduction of the species. Both Medawar and Williams emphasized this point, although a more precise version of the argument was given by Hamilton (1966). We shall review here only the simplified situation where the population in question is in a steady state, i.e. has a stable age distribution and constant size; the reader interested in a more general treatment may refer to Hamilton's paper.

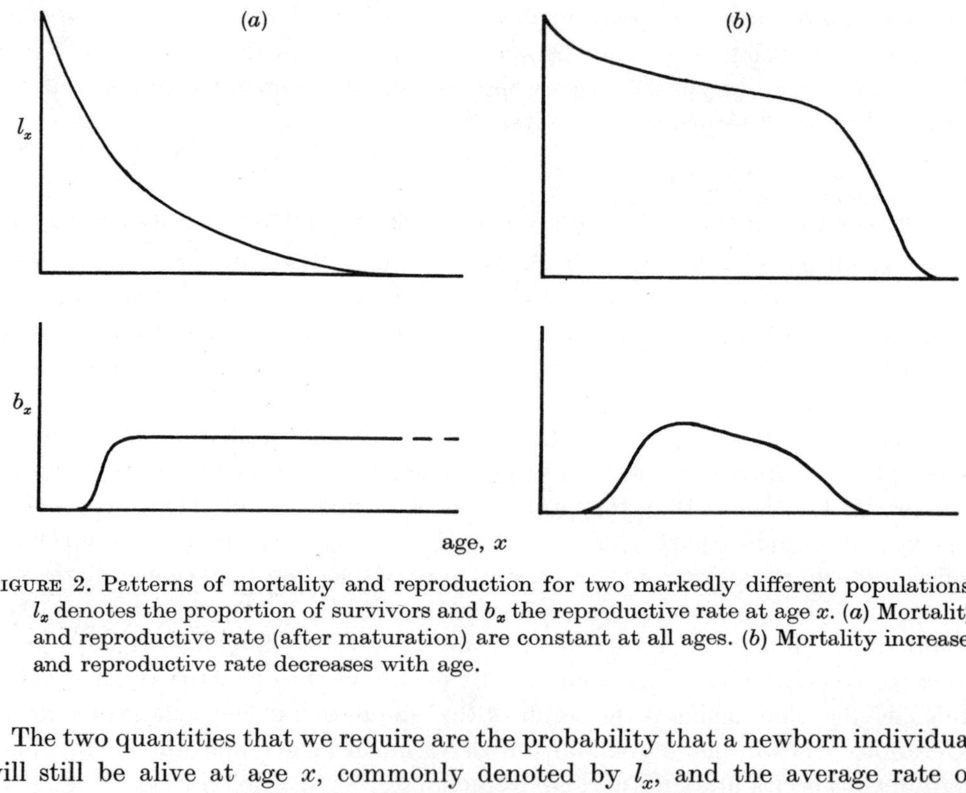

FIGURE 2. Patterns of mortality and reproduction for two markedly different populations; l_x denotes the proportion of survivors and b_x the reproductive rate at age x. (*a*) Mortality and reproductive rate (after maturation) are constant at all ages. (*b*) Mortality increases and reproductive rate decreases with age.

The two quantities that we require are the probability that a newborn individual will still be alive at age x, commonly denoted by l_x, and the average rate of reproduction, b_x, of individuals aged x. Two contrasting situations are illustrated in figure 2. Figure 2*a* shows a population with no ageing; $l_x = \mathrm{e}^{-\mu x}$ where μ is a constant death rate, and the rate of reproduction is constant at all ages. Figure 2*b* shows a population that does age, as indicated by the increase in mortality and decrease in reproductive rate. (We ignore periodic effects such as seasonality and the complications of sexual reproduction since these have no radical influence on the general argument.) Multiplying l_x and b_x together gives a quantity $p_x = l_x b_x$ that represents, for a newborn individual, the *expected* rate of reproduction at age x, taking into account the probability that it may die before it reaches age x. The integral $\int_0^\infty p_x \,\mathrm{d}x$ is the total expected reproduction for a newborn individual. This equals unity, by definition, in the steady state situation.

For the two populations shown in figure 2, the relations between p_x and x are similar, despite the major differences in b_x and l_x, and both approximate the broken curve in figure 3. Up to the age of first reproduction, p_x is zero. It rises thereafter to a maximum and falls away again to zero. The solid curve indicates the quantity $s_x = \int_x^\infty p_z \, dz = 1 - \int_0^x p_z \, dz$. This is the fraction of the total expected reproduction (area under the broken curve) that occurs after any given age x, and its rate

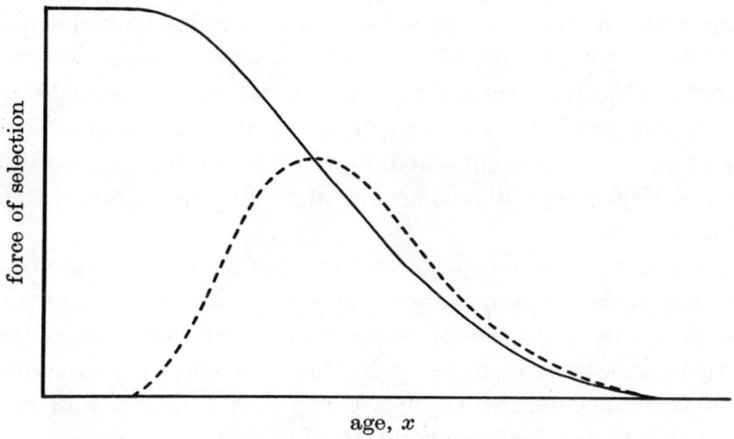

FIGURE 3. The broken curve represents the quantity $p_x = l_x b_x$ (see text) for a typical population; this equals the expected rate of reproduction at age x for a newborn individual, taking account of the possibility of prior mortality. The area under the broken curve is the total expectation of reproduction. The solid curve, representing the remaining fraction of the total expectation of reproduction at any age, is a measure of the force of natural selection acting at that age.

of fall indicates the expected rate at which the total expectation of reproduction will be used up through the lifespan. Natural selection operates through reproduction and therefore acts more forcefully on those portions of the lifespan where the remaining fraction of the total expected reproduction is greatest. The quantity s_x provides us with a quantitative measure of how the force of natural selection declines steadily after the start of reproduction. As Hamilton (1966) was careful to demonstrate, this quantity should not be confused with Fisher's (1930) definition of reproductive value which is equivalent to the expected reproduction of an individual of age x and which does not allow for the chance of prior mortality. Fisher's reproductive value is not a valid measure of the force of natural selection.

The importance of the work of Medawar, Williams and Hamilton is the clear elucidation of the effect of age on the force of natural selection. This cannot fail to have relevance to the evolution of ageing since it shows how selective control over the later portions of the lifespan must progressively tail away. There is, however, a serious logical objection to explaining the origin of ageing by the theory of accumulating late-expressed deleterious genes. This was first pointed out by

Kirkwood (1977) and has been reiterated by Sacher (1978). The objection is that the concept of 'late expression' itself implies the prior existence of adult age-related physiological processes. An organism that does not age is, in effect, in a steady state, where the physiology of chronologically young and old animals is the same. In a situation like this it is difficult, if not impossible, to see how genes could measure the passage of time and have early or late expression. However, if ageing already existed within a population, there would certainly be the potential for precession or recession of good and bad genes since the ageing process itself could define the time scale. In this case, however, the accumulation of deleterious genes at the end of the lifespan would be a reinforcing consequence, not the cause, of ageing. The same objection applies equally to the pleiotropic gene hypothesis, except where the bad gene effects are simply the cumulative result of the continued expression of a gene which has outlasted its useful function. In general, the risk of circularity in the Medawar–Williams argument needs more careful attention than it has hitherto received.

So far, the theories we have considered have not been at all specific about the possible molecular basis of ageing. Indeed, they have suggested that a multitude of processes may be involved. In the following sections, we shall suggest that there is good reason to believe that ageing may be due primarily to a breakdown in the accuracy of macromolecular synthesis within cells and we shall consider the implications of this theory for the evolution of longevity.

Errors in information transfer between macromolecules

The basis of terrestrial life is the replication and translation into protein of genetic information stored in polynucleotide sequences within cells. Orgel (1963) pointed out that the molecular apparatus responsible for transcription and translation contained the potential for a lethal feedback of errors since it itself was largely made up of proteins. Occasional errors in the synthesis of translational proteins could lead to the creation of molecules which retained activity but which had altered specificity and so would generate further errors. Several authors have generalized the original concept to include errors of DNA replication, since proteins (DNA polymerases) are crucially involved in the accuracy of this process (Woolhouse 1969; Lewis & Holliday 1970; Holliday & Tarrant 1972; Orgel 1973; Burnet 1974). A progressive deterioration in the accuracy of macromolecular synthesis has been proposed as a possible mechanism for cellular ageing, which in turn could be responsible for at least part of the decline in vitality of ageing organisms (Medvedev 1962; Orgel 1963, 1973).

Orgel originally suggested that a build-up of errors leading to what he termed an 'error catastrophe', when the cell could no longer function, would be the inevitable consequence of errors within the cell. If this were the case, however, it is hard to see how, even through the action of cellular selection, life could have avoided total extinction, or could ever have started at all. In an important revision of his original paper, Orgel (1970) described a simple model which showed how a cell could either maintain a stable low level of errors or experience error catastrophe. Theoretical work by Hoffman (1974) reaffirmed these alternatives, but suggested that error catastrophe would be very unlikely in present-day organisms.

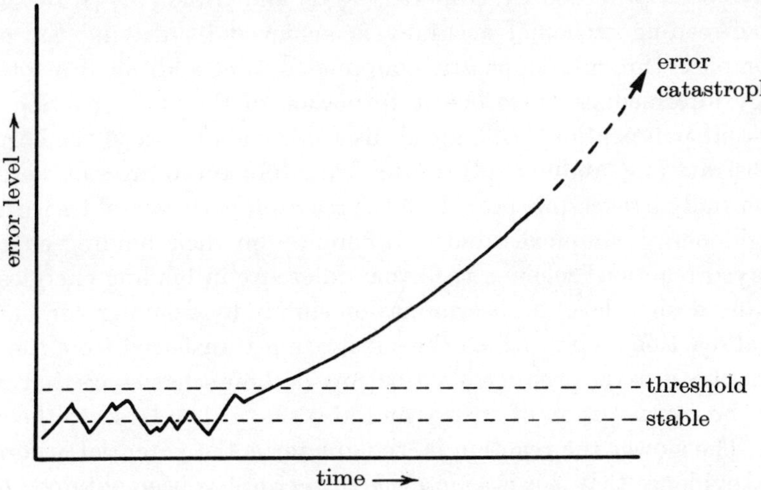

FIGURE 4. The level of errors in macromolecular synthesis within a cell is normally maintained close to the stable value. Random fluctuations in the error level may occasionally take the cell across the threshold above which unrestrained error propagation occurs, leading ultimately, to a lethal error catastrophe.

However, Hoffman's model contained a seriously restrictive assumption and we have shown that a more realistic generalization of it makes error propagation more plausible (Kirkwood & Holliday 1975). All theoretical studies of Orgel's error theory (for a review, see Kirkwood 1979) concur in the major conclusion that, depending on the amount of error feedback, the level of errors will stabilize at some low value or will increase steadily until, perhaps after several cell generations, it reaches a lethal point. Even for a cell with stable translation, there will be a threshold error level above which the cell becomes committed to error catastrophe. Random fluctuations in the error level may result in the cell crossing this threshold from the stable to the unstable state (figure 4). The risk that this happens will depend on the distance between the stable and threshold error levels relative to the magnitude of the error level fluctuations which, in turn, depends on the specificity of macromolecular synthesis and on the relative activities and degradation rates of erroneous and correct products (Kirkwood & Holliday 1975). In principle, both of these factors can be controlled by natural selection through the evolution of proof-reading and scavenging enzymes. For example, it is well known that the accuracies of protein synthesis and of DNA replication considerably exceed what would be predicted on purely stereochemical grounds. In the case of

DNA replication, it has been shown that mismatched bases are removed by the exonuclease activity of DNA polymerase before further bases can be incorporated in the growing strand (see Kornberg 1974). This is an energy consuming process, since a certain proportion of nucleoside triphosphates are converted to nucleoside monophosphates. Generalized schemes for reducing errors in macromolecular synthesis have been proposed by Hopfield (1974) and Ninio (1975). In Hopfield's 'kinetic proof-reading' scheme, accuracy is achieved by driving the enzyme–substrate complex, through enzymatic coupling to some additional reaction, into a high energy intermediate state before formation of the main reaction product (e.g. the nascent polypeptide). Enhanced discrimination between the correct and incorrect substrate (e.g. amino acid) results from differential rates of dissociation, along an essentially irreversible branch of the reaction pathway, of the correct and incorrect high-energy complexes due to differences in their binding energies. In Ninio's 'delayed reaction' scheme, the same difference in binding energies is used to generate any desired level of discrimination simply by delaying, for sufficiently long, the final reaction step in which the substrate is transferred from the enzyme substrate complex into the product. Two features of these schemes are that accuracy is related to the consumption of energy and also that it is related to the speed of the reaction. The slower the reaction is, the greater is the potential accuracy and experimental evidence that this is so in *Escherichia coli* has been obtained (Galas & Branscomb 1976). Although it may seem that the amount of energy consumed in proof-reading is relatively trivial compared with the overall energy needed to synthesize the components of macromolecules, it should be pointed out that the whole apparatus of protein synthesis, including the complexity of the ribosome, may well be crucial for accuracy. Moreover, a large amount of energy is dissipated in the continual turnover of proteins in higher organisms. It is possible that the physiological function of this process is to remove abnormal and potentially dangerous molecules.

Validation of Orgel's error propagation hypothesis requires the demonstration that feedback of translational errors does occur. Several workers have investigated this by artificially raising the level of errors in *E. coli* through the addition of low concentrations of streptomycin, an error-inducing antibiotic, to the growth medium. Branscomb & Galas (1975) reported that after addition of the drug they found a progressive increase in the level of erroneous protein and a delayed progressive decrease in growth rate of the cells. They concluded that error feedback does occur in *E. coli* and that the cells can be induced to experience error catastrophe. These results have been confirmed and extended by Rosenberger & Holliday (1979). However, Edelmann & Gallant (1977) performed similar experiments and found that the initial rise in error frequency after addition of streptomycin subsequently declined and that the error frequency reached a limiting value, reverting to normal on removal of the drug. In fact, these results are not as incompatible as they first appear and may be explained in terms of different threshold error levels in the two experimental systems (Kirkwood 1979). Interestingly, Edelmann & Gallant's own calculation suggested a significant amount of error feedback, even though the cells remained stable.

Evolution of ageing

In evolutionary terms, errors in macromolecular synthesis are both good and bad since too many of them reduce reproductive efficiency by generating too many mutations, while too few restrict the scope for natural selection to adapt a species to environmental change. Clearly the optimal error level for any organism is a compromise related to its ecological niche. Kirkwood (1977) suggested an additional reason, providing the basis of a new theory of the evolution of ageing, why error levels should be a compromise between high and low. Poor control over the accuracy of macromolecular synthesis renders a cell more vulnerable to the risk of error catastrophe. Improved control through proof-reading diminishes this risk but only at the expense of increased consumption of energy and slower growth.

For a single-celled organism the consequences of crossing the threshold error level are indeed catastrophic since its reproductive potential is then eliminated. (The cell may subsequently divide but all progeny are committed to error catastrophe.) We therefore expect selection to have favoured a high degree of translational stability, as seems to be so in *E. coli* (Edelmann & Gallant 1977), at the expense of some increase in energy requirement or generation time. In multicellular organisms the situation is somewhat different. The integrity of macromolecular synthesis must be carefully preserved in germ-line cells since reproduction and long-term survival depend on it; this can be ensured either by proof-reading or scavenging erroneous molecules or by cellular selection, perhaps aided by a special 'suicide' protein that kills the cell if the error level crosses the threshold and begins to rise (Orgel 1963; Holliday 1974). Somatic cells, however, play only an indirect part in long-term gene survival since each individual soma is certain to die sooner or later of some accidental cause. We have already noted for most species longevity is determined more by the rate of accidental mortality than by any intrinsic ageing. There is, therefore, no point in an organism investing more resources in somatic maintenance than is necessary for it to last sufficiently long in adequate reproductive condition to realize its normal expectation of life. In particular, by reducing control over the accuracy of somatic macromolecular synthesis to the minimum viable level, an organism could either reduce its energy needs or accelerate its reproduction or both. The penalty for this saving would be a progressive increase through life in the proportion of cells committed to error catastrophe, leading ultimately to a fatal breakdown of homeostasis, but this would occur sufficiently late and affect sufficiently few individuals not to detract from the selective benefit of the strategy. This offers a simple and non-circular explanation of the evolution of ageing (Kirkwood 1977). The strategy is analogous to that used in the manufacture of disposable goods, which have a short expected life and are therefore made as cheaply as possible; we shall subsequently refer to it as the 'disposable soma' strategy or theory.

The disposable soma theory is, in a sense, a special case of Williams's (1957) pleiotropic gene hypothesis, the gene in question controlling the switch to reduced accuracy in somatic cells. The good effect of the gene is the reduced investment of resources in the soma, while the bad effect is ultimate somatic disintegration, or ageing. (The concept may be generalized to include somatic repair and renewal systems other than those involved in intracellular macromolecular synthesis, but these probably evolved later and we shall not consider them in detail here.) Maximum benefit is obtained by switching as early as possible, at or around the time of differentiation of somatic cells from the germ line. The theory applies strictly only where differentiation of soma from the germ line is rigid and normally irreversible. When differentiation is more plastic, for example in plants, some modification is necessary. Ageing in such organisms is in any case generally less well defined.

Evolution of longevity

The life history of a species is described by a set of co-adapted traits that together define its pattern of mortality and fecundity (Stearns 1976). As described above, these same traits determine the rate of decline of the force of natural selection with age. Williams (1957) and Hamilton (1966) presented convincing arguments that this rate of decline should be closely associated with the rate of ageing, so that any specific theory of the evolution of ageing and longevity should conform with this requirement. In this section we describe how the disposable soma theory meets this requirement in the case of iteroparous species (i.e. those that reproduce repeatedly).

In general, high mortality is associated with rapid decline in the force of selection, and low mortality with a slower decline. We should therefore expect species subject to heavy predation or other hazards to age more rapidly than those with few predators. This expectation is obviously satisfied if we compare widely differing species, for example small rodents with large mammals, but striking comparisons can also be made even among small mammals. The mouse, *Mus musculus*, has many predators and lives to a maximum of about 3 years, while the little brown bat, *Myotis lucifugus*, which lives in a special environment and can evade many predators by flight, has an extended lifespan, probably as long as 24 years (Griffin & Hitchcock 1967). In general, adaptation improves fitness either by reducing mortality or by increasing the reproductive rate, so we also expect a high risk of mortality and rapid ageing to be associated with large, frequent litters. Female mice in the wild produce an average of approximately 45 offspring per year, compared with only 1 for *M. lucifugus* (Asdell 1964).

These differences between mice and bats are readily explained by the disposable soma theory. A species occupying an ecological niche where predators and other hazards are common ought not to invest too heavily in an individual soma, which as a consequence will age rapidly and be short-lived, but should concentrate instead on producing as many offspring as cheaply and as quickly as it can. On the other hand, a species gaining an adaptation such as flight that enables it to colonize a more secure niche will alter the curve of its life-table in the direction *b* to *c* in figure 1 so that a greater proportion of the population experiences the harmful consequences of ageing. The fact that the lifespan is then limited by

TABLE 1. EFFECT OF MORTALITY AND LIFESPAN ON REPRODUCTIVE FITNESS (EXPRESSED AS THE MULTIPLICATIVE GROWTH RATE IN POPULATION SIZE PER DECADE) FOR A SPECIES WITH A REPRODUCTIVE PATTERN ROUGHLY SIMILAR TO THAT OF THE INDIAN ELEPHANT

(Females are assumed to reach maturity at 10 years and to produce one offspring then and at 5-year intervals thereafter until death. Death is assumed to occur at the specified maximum age if an individual survives until then; otherwise, the mortality is constant at all ages. It is assumed that the sex ratio is 50:50 and the population growth rate is calculated for females only, counting each offspring as representing, on average, 0.5 female.)

mortality (% per decade) / maximum lifespan (years)	10	20	30	40	50
20	1.18	1.05	< 1	< 1	< 1
40	1.58	1.40	1.23	1.05	< 1
60	1.65	1.47	1.28	1.10	< 1
80	1.67	1.48	1.30	1.11	< 1
∞	1.69	1.50	1.31	1.12	< 1

ageing means that the full benefit of the adaptation cannot be realized, resulting in selective pressure to extend the maximum lifespan. It is instructive to consider the effect of longevity on the fitness of a species that reproduces slowly. The female Indian elephant becomes sexually mature at 10–15 years of age, has a gestation period of about 620 days and produces only one offspring at a time. Not more than two offspring are produced per decade (Asdell 1964). Table 1 shows the interrelation of mortality (assumed constant with age), maximum lifespan (beyond which survivors are assumed to die 'of old age') and fitness (expressed as the multiplicative growth rate e^m where m is the Malthusian parameter) for a population with a reproductive pattern like that of the elephant. It can be seen that decreasing the mortality increases the negative impact of ageing on fitness, while for any given level of mortality the advantage of lengthening the lifespan progressively diminishes. In terms of the disposable soma theory, a decrease in mortality alters the optimal error level in somatic cells in favour of increased accuracy. This would postpone ageing at the cost of retardation of growth or increased food requirements so the net effect on the lifetable would be to stretch out the time scale. Such a process would fit with the evolution of greater brain size and higher intelligence (Sacher 1978), requiring extended infancy and protracted parental care for the education of the young.

Williams (1957) observed that ageing should always be a generalized phenomenon, and never due largely to changes in a single physiological system. He used the conclusion that many physiological and anatomical changes would accumulate during ageing to dispel the notion of the 'fountain of youth'. His argument does not, however, rule out the possibility of a primary pacemaker system that controls the overall rate of ageing and permits relatively rapid evolutionary change in longevity. If the disposable soma theory is correct, the mechanisms of intracellular error regulation may prove to be such a system. It is much easier to understand how natural selection could act to increase the fidelity of synthesis of macromolecules than it is to see how it could remove a series of more or less synchronized late acting deleterious genes, any one of which would result in death.

Developmental programmes and ageing

So far we have not discussed the possibility that ageing is in some way programmed. Although this theory is popular among gerontologists, we are not aware of any serious attempt to explain how a programme could evolve. Nevertheless, once ageing has evolved as a result of reduced accuracy of macromolecular synthesis in somatic cells, then developmental changes could be superimposed on this as a secondary effect on the basic process of ageing. If an organism can in any case only survive for a given period of time, there is little point in it having components which are capable of lasting for a much longer period. For example the teeth of the horse, which wear down with time, usually last long enough to allow the animal to feed until senescence, but some old animals may well die through being unable to feed. Should we, therefore, say that the wearing out of teeth is a case of programmed ageing? In general the disposable soma theory, as well as that of Medawar and Williams, predicts that senescence will be associated with more or less synchronized changes in many tissues. Some of these may be due to error accumulation or the failure to renew cells; others may be due to developmental processes. A clear case of a programmed age-related developmental change is the menopause in the human female. Since this appears to be restricted to our own species, it seems likely that it is associated with some specific aspect of human reproduction or development. One possibility is that the very slow development of children and the degree of maternal care involved makes it advantageous for a woman to cease reproduction while she is bringing up her later offspring. Another, which is mentioned by Medawar (1952) is that genes for 'grandmotherly indulgence' may exist. It is obvious that post-menopausal women play an active role in many societies in caring for their grandchildren. Although infertile, they are increasing the probability that some of their genes will survive, in other words their inclusive fitness.

Apart from age-related developmental changes, there are scattered examples of ageing that is specifically programmed. One which is often cited is the rapid death of the migratory salmon after spawning. Another is the death of the female octopus a few weeks after her eggs have hatched. There is evidence that both are triggered by hormonal changes. In the octopus, it has been specifically shown that the optic gland is responsible. Removal of the gland allows the female to survive very much longer than she would otherwise do (Wodinsky 1977). One possible explanation of these situations is the female puts so many of her resources into producing offspring, that she is in any case unlikely to survive to produce a second

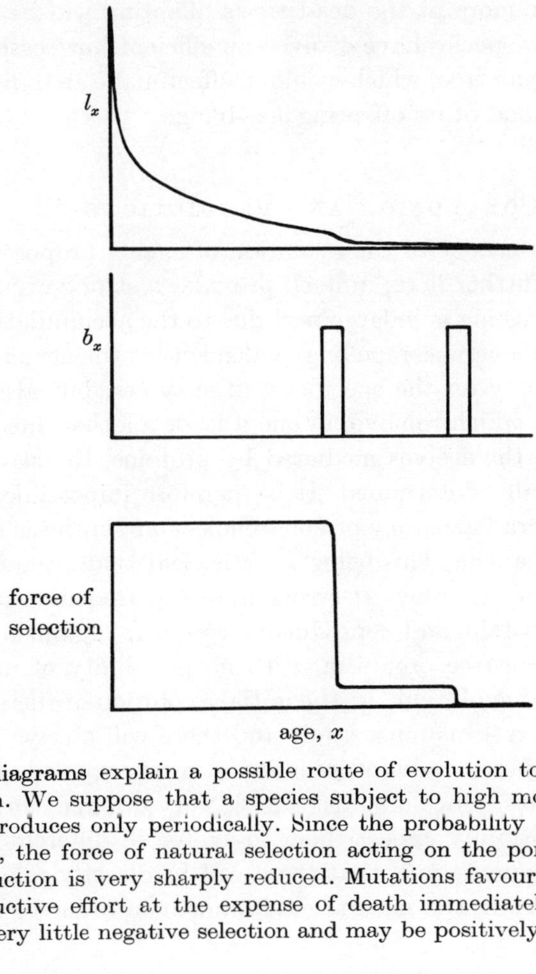

FIGURE 5. The diagrams explain a possible route of evolution to semelparous (once only) reproduction. We suppose that a species subject to high mortality, especially during infancy, reproduces only periodically. Since the probability of surviving to reproduce twice is low, the force of natural selection acting on the portion of the lifespan after first reproduction is very sharply reduced. Mutations favouring greater success in the first reproductive effort at the expense of death immediately after reproduction will encounter very little negative selection and may be positively adaptive.

brood. This is illustrated in figure 5. If the likelihood of survival past the first act of reproduction is very low, then there will be little selection against developmental changes that result in death after reproduction. This is fairly convincing in the salmon, since to achieve a second reproductive cycle, the hazards associated with migration have to be circumvented. A second possibility is that Weismann's argument holds good for certain species in particular environments. For instance, the death of a parent may increase the probability of the offspring's surviving. This might be due to a temporary increase in the richness of the local microflora or fauna after the decay of the adult, or just conceivably, in the female octopus, the vacation of territory. Forest trees may provide an example of the validity of Weismann's theory. Trees grow in size indefinitely and their longevity is largely determined by the limitations of their mechanical strength. The forest canopy effectively prevents any growth of seedlings, until a tree falls. There then is intense competition to fill the space and since in many species little seed dispersal occurs, it is likely that one or more of the dead tree's offspring will be successful. It is revealing that very few species have evolved an efficient buttressing system (such as that seen in the banyan tree) which would in effect make an individual immortal and restrict the likelihood of its offspring surviving.

Conclusion and predictions

The disposable soma theory of the evolution of ageing proposed by Kirkwood (1977) and developed further here, in itself provides a strong argument in favour of the hypothesis that ageing is in large part due to the accumulation of defects in macromolecules. There is now compelling evidence that the accuracy of synthesis of these depends not only on the specificity of enzymes, but also on a series of additional mechanisms which remove, in one way or another, incorrect products. These mechanisms are themselves mediated by proteins, the structure of which are, of course, genetically determined. It is therefore impossible to dispute the conclusion that the overall accuracy of macromolecular synthesis is dependent on the genotype of the organism. This being so, it is clear that organisms will evolve to an optimum level of accuracy. If errors were too frequent, it would be impossible to develop, sustain and reproduce a complex organism. On the other hand, a completely error-free organism, with no possibility of mutation, would no longer be capable of evolution. In the initial evolution of increasing accuracy the law of diminishing returns must apply and there will always be an optimum point where there is no further advantage in investing energy and cellular resources to reduce the error levels. Since organisms have very different environments and life styles we would certainly expect them to evolve to different optimum error levels, and our basic hypothesis is that these will be directly related to the time the species must survive in the natural environment in order to reproduce itself successfully.

The theory we have described makes two important predictions, both of which can be tested experimentally. First, we would predict that germ-line cells synthesize macromolecules with greater accuracy than somatic cells. So far, no direct comparison of the accuracy of protein synthesis in these two types of cell have been made, but measurements of this type should soon be possible. Although there is considerable information about frequency of defects in DNA (mutations) in the germ line of various species, surprisingly, there is remarkably little information about somatic mutation rate *in vivo*. As Burnet (1974) and others have emphasized, there is a real need for much more research on the frequencies of somatic mutations. It is, of course, possible that accuracy is much the same in somatic cells and germ-line cells, but that there are special mechanisms which eliminate any gametes, or their precursors, that have accumulated a significant number of errors. This is also open to experimental investigation.

The second basic prediction is that the accuracy of macromolecular synthesis in somatic cells is directly related to the lifespan of the species. This could also be tested at the level of protein, RNA or DNA synthesis, but so far no appropriate experiments have been done. There is, however, one very important series of observations concerning the capacity for DNA repair in different mammals. Hart & Setlow (1974, and personal communication) cultured fibroblasts from a number of species, irradiated them with a constant dose of ultraviolet light and then measured the amount of repair synthesis in nuclei outside the S phase. (The amount of repair is a measure of the proportion of pyrimidine dimers removed, rather than the length of the repair synthesis patch.) The striking result is that longer lived mammals are much more efficient at excision repair than short-lived ones. The same effect is seen for two fairly closely related rodents, *Mus musculus* and *Peromyscus leucopus*, whose maximum lifespans differ by nearly threefold. The longer lived species, *P. leucopus*, has a more efficient excision repair mechanism (Sacher & Hart 1978). Chemical defects occur spontaneously in DNA, and their enzymic repair is one of the mechanisms used to preserve the integrity of these molecules (see, for example, Lindahl *et al.* 1978). The results of Hart, Setlow and Sacher therefore support the disposable soma theory and emphasize the need for further comparative studies of the accuracy and efficiency of macromolecular synthesis and repair.

We thank L. E. Orgel, F.R.S., and J. E. Pugh for helpful discussions.

References (Kirkwood & Holliday)

Asdell, S. A. 1964 *Patterns of mammalian reproduction*, 2nd edn. London: Constable.

Branscomb, E. W. & Galas, D. J. 1975 Progressive decrease in protein synthesis accuracy induced by streptomycin in *Escherichia coli*. *Nature, Lond.* **254**, 161–163.

Burnet, F. M. 1974 *Intrinsic mutagenesis: a genetic approach to ageing*. New York: J. Wiley.

Comfort, A. 1964 *Ageing: the biology of senescence*. London: Routledge & Kegan Paul.

Edelmann, P. & Gallant, J. 1977 On the translational theory of ageing. *Proc. natn. Acad. Sci. U.S.A.* **74**, 3396–3398.

Fisher, R. A. 1930 *The genetical theory of natural selection*. Oxford: Clarendon Press.

Galas, D. J. & Branscomb, E. W. 1976 Ribosomes slowed by mutation to streptomycin resistance. *Nature, Lond.* **262**, 617–619.

Griffin, D. R. & Hitchcock, H. B. 1967 Probable 24 year longevity records for *Myotis lucifugus*. *J. Mammal.* **46**, 332.

Hamilton, W. D. 1966 The moulding of senescence by natural selection. *J. theor. Biol.* **12**, 12–45.

Hart, R. W. & Setlow, R. B. 1974 Correlation between deoxyribonucleic acid excision – repair and lifespan in a number of mammalian species. *Proc. natn. Acad. Sci. U.S.A.* **71**, 2169–2173.

Hoffman, G. W. 1974 On the origin of the genetic code and the stability of the translation apparatus. *J. molec. Biol.* **86**, 349–362.

Holliday, R. 1974 Growth and death of diploid and transformed human fibroblasts. *Fedn Proc. Fedn Am. Socs exp. Biol.* **34**, 51–55.
Holliday, R. & Tarrant, G. M. 1972 Altered enzymes in ageing human fibroblasts. *Nature, Lond.* **238**, 26–30.
Hopfield, J. J. 1974 Kinetic proofreading: a new mechanism for reducing errors in biosynthetic processes requiring high specificity. *Proc. natn. Acad. Sci. U.S.A.* **71**, 4135–4139.
Kirkwood, T. B. L. 1977 Evolution of ageing. *Nature, Lond.* **270**, 301–304.
Kirkwood, T. B. L. 1979 Error propagation in intracellular information transfer. Submitted to *J. theor. Biol.*
Kirkwood, T. B. L. & Holliday, R. 1975 The stability of the translation process. *J. molec. Biol.* **97**, 257–265.
Kornberg, A. 1974 *DNA synthesis*. San Francisco: W. H. Freeman.
Lack, D. 1954 *The natural regulation of animal numbers*. Oxford: Clarendon Press.
Lewis, C. M. & Holliday, R. 1970 Mistranslation and ageing in *Neurospora*. *Nature, Lond.* **228**, 877–880.
Lindahl, T., Karran, P. & Riazuddin, S. 1978 DNA glycosylases of *Escherichia coli*. In *DNA repair mechanisms* (ed. P. C. Hanawalt, E. C. Friedberg & C. F. Fox), pp. 179–186. New York: Academic Press.
Medawar, P. B. 1952 *An unsolved problem in biology*. London: H. K. Lewis.
Medvedev, Zh. A. 1962 Ageing at the molecular level. In *Biological aspects of ageing* (ed. N. W. Shock), pp. 255–266. New York: Columbia University Press.
Ninio, J. 1975 Kinetic amplification of enzyme discrimination. *Biochimie* **57**, 587–595.
Orgel, L. E. 1963 The maintenance of the accuracy of protein synthesis and its relevance to ageing. *Proc. natn. Acad. Sci. U.S.A.* **49**, 517–521.
Orgel, L. E. 1970 The maintenance of the accuracy of protein synthesis and its relevance to ageing: a correction. *Proc. natn. Acad. Sci. U.S.A.* **67**, 1476.
Orgel, L. E. 1973 Ageing of clones of mammalian cells. *Nature, Lond.* **243**, 441–445.
Rosenberger, R. F. & Holliday, R. 1979 Streptomycin induced protein errors and limited lifespan of *Escherichia coli*. In preparation.
Sacher, G. A. 1978 Evolution of longevity and survival characteristics in mammals. In *The genetics of ageing* (ed. E. L. Schneider), pp. 151–167. New York: Plenum.
Sacher, G. A. & Hart, R. W. 1978 Longevity, ageing and comparative cellular and molecular biology of the house mouse, *Mus musculus*, and the white-footed mouse *Peromyscus leucopus*. In *Genetic effects on ageing* (*Birth defects: original article series* no. 14) (ed. D. Bergsma & D. E. Harrison), vol. 1, pp. 71–96. New York: Alan R. Liss.
Stearns, S. C. 1976 Life-history tactics: a review of the ideas. *Q. Rev. Biol.* **51**, 3–46.
Weismann, A. 1891 In *Weismann on heredity* (ed. E. B. Poulton, S. Schönland & A. E. Shipley), 2nd edn, pp. 23–42. Oxford University Press.
Williams, G. C. 1957 Pleiotropy, natural selection and the evolution of senescence. *Evolution* **11**, 398–411.
Wodinsky, J. 1977 Hormonal inhibition of feeding and death in *Octopus:* control by optic gland secretion. *Science, N.Y.* **198**, 948–951.
Woolhouse, H. W. 1969 DNA polymerase, genetic variation and determination of the lifespan. In *Proc. 8th Int. Congr. Geront.*, vol. 1, pp. 162–166. Washington, D.C.: Federation of American Societies for Experimental Biology.

Proc. R. Soc. Lond. B **205**, 547–565 (1979)
Printed in Great Britain

Comparison and adaptation

By T. H. Clutton-Brock† and P. H. Harvey‡
† *King's College Research Centre, Cambridge, U.K.*
‡ *School of Biological Sciences, University of Sussex, Falmer, Brighton BN1 9QG, U.K.*

It has sometimes been suggested that the term adaptation should be reserved for differences with a known genetic basis. We argue that adaptation should be defined by its effects rather than by its causes as any difference between two phenotypic traits (or trait complexes) which increases the inclusive fitness of its carrier. This definition implies that some adaptations may arise by means other than natural selection. It is particularly important to bear this in mind when behavioural traits are considered.

Critics of the 'adaptationist programme' have suggested that an important objection to many adaptive explanations is that they rely on *ad-hoc* arguments concerning the function of previously observed differences. We suggest that this is a less important problem (because evolutionary explanations generally claim some sort of generality and are therefore testable) than the difficulties arising from confounding variables. These are more widespread and more subtle than is generally appreciated.

Not all differences between organisms are directly adapted to ecological variation. The form of particular traits usually constrains the form or value that other traits can take, presenting several obstacles to attempts to relate variation in morphological or behavioural characteristics directly to environmental differences. We describe some of the repercussions of differences in body size among vertebrates and ways in which these can be allowed for.

In addition, a variety of evolutionary processes can produce non-adaptive differences between organisms. One way of distinguishing between these and adaptations is to investigate adaptive trends in phylogenetically different groups of species.

1. Concepts of adaptation

Most adaptive explanations for traits or trait complexes are founded initially on comparison. We observe that giraffe have long necks while buffalo have short ones; we note that giraffe spend much of their time browsing on tree tops while buffalo graze at ground level; and we hypothesize that the giraffes' long necks are an adaptation to feeding on tree tops.

Application of this basic methodology, which was extensively employed by

Darwin (1859, 1871) and has been used by evolutionary biologists ever since, faces at least three important problems. First, most explanations based on comparative evidence are initially, like the example above, *post hoc*: explanations are fitted to previously observed facts. In some cases, critics have even complained that they are commonly *ad hoc*, claiming no generality beyond the particular set of facts which they seek to explain, and consequently untestable (Lewontin 1979).

Secondly, they face problems resulting from confounding variables. Not only are behavioural, physiological and morphological traits closely associated with each other, but so are many environmental variables. Neck length, for example, probably affects both the ability to feed at different levels and the ability to detect approaching predators. How can we determine whether the long necks of giraffe are an adaptation connected with tree tops rather than predators?

Thirdly, a proportion of differences between species may not be directly adapted to any ecological differences. Not only may certain adaptations constrain the form of other aspects of the species' morphology, physiology or behaviour, but a variety of evolutionary mechanisms may produce non-adaptive differences between species (Lewontin 1978, 1979; Gould 1978). One of the most important of these is the tendency for different species to occupy alternative adaptive peaks, producing different responses to similar ecological problems (Lewontin 1978). Sloths, for example, have managed to utilize tree tops by means of adaptations other than long necks.

Before discussing these three problems it is important to clarify our definition of adaptation. We define adaptation as a difference between two phenotypic traits (or complexes of traits) which increases the inclusive fitness (Hamilton 1964*a*, *b*) of its carrier. In many cases, direct assessment of the effects of a difference in reproductive success is unattainable and we rely on evidence that the trait shows a closely designed fit to some problem presented by the animal's environment (see Williams 1966). However, we should not allow this to obscure the fact that adaptation is relative: a trait is adaptive compared with an alternative form or value of the same characteristic or even to its absence (Hinde 1975).

Our definition of adaptation also carries the implication that all adaptations need not necessarily have genetic causes. The term is generally applied to phenotypic differences and, particularly where behavioural differences are concerned, we sometimes cannot be certain that a particular difference has a genetic basis. For example, across primate species, groups occupy home ranges of very different sizes (Clutton-Brock & Harvey 1977*a*). These differences are related to food abundance and habitat, and are obviously adaptive: those members of a group that adopted a home range that provided inadequate supplies of food would show low reproductive success. However, home range size may vary between neighbouring groups and between populations of the same species by several orders of magnitude (Clutton-Brock 1977) and it seems entirely possible that individuals of different species may learn to adopt the size of home range that meets their energetic requirements most satisfactorily. The same argument applies to many differences in social behaviour where cultural transmission as well as individual learning may be involved (Kummer 1971; Carpenter 1973).

We have sometimes heard it argued that the concept of adaptation should be restricted to differences which have a known genetic origin. In the case of learned differences in behaviour – so the argument runs – it is not the phenotypic difference that is adaptive but the inherited potential for learning it.

Let us erect a straw biologist who adopts this position. A hypothetical field worker has observed two populations of baboons, the first of which is subject to heavy predator pressure while the second is not. Individuals of the first population consistently sleep in palm trees, thereby reducing their chances of being eaten, while those belonging to the second population do not. Suppose, too, that the difference in sleeping behaviour between the two populations can be shown to be learned, perhaps by exchanging individuals between them. According to our definition (see above), the difference in sleeping behaviour would be an adaptation to predator pressure, despite the fact that it had no genetic basis. In contrast, our straw biologist would argue that it is not the phenotypic difference in behaviour but the (presumably) inherited general learning ability that permits the baboons to learn to sleep in palm trees which is adaptive.

We should not disagree that the ability to learn to sleep in the best place (as against a set of inflexible genetic instructions constraining baboons to sleep in particular sites) may be adaptive. However, we should argue that the phenotypic difference in sleeping behaviour is also an adaptation and that it is impractical to attempt to restrict the concept to differences of known genetic origin. Even if it were possible to determine which behavioural differences were the consequences of differences in learning potential, we should still be unsure that these were caused by genetic differences since there is ample evidence that the ability to learn can itself be affected by experience (Hinde 1970). Moreover, in the great majority of cases where both experience and inherited factors are involved in ontogeny, our straw biologist's position would logically require him to identify those components of the difference that had a genetic basis and those that did not, regarding only the former as adaptive. The difficulty of separating clearly between inherited and acquired characteristics was well appreciated by Darwin (1859, chapter 4): 'When a variation is of the slightest use to any being, we cannot tell how much to attribute to the cumulative action of natural selection, and how much to the definite action of the conditions of life. Thus it is well known to furriers that animals of the same species have thicker and better fur the further north they live; but who can tell how much of this difference may be due to the warmest clad individuals having been favoured and preserved during many generations and how much to the action of the severe climate?'

As we have already suggested, we believe that the solution to this problem is to define adaptation by its effects rather than its causes. The fact that certain adaptations may have a largely non-genetic origin in no way weakens the concept, as has sometimes been implied. Indeed, the presence of mechanisms that can modify behaviour and morphology in an adaptive fashion within the lifespan of an individual indicates that organisms can be more closely fitted to their environments than if adaptation could only be produced by genetic change (Braestrup 1971). It does, however, imply that adaptive differences can arise through a variety of processes other than natural selection (see Tinbergen 1972; Gould 1978).

2. *Ad-hoc* explanations and the problems of confounding variables

As critics have pointed out, the logical structure of many adaptive arguments is frequently weak: a difference in behaviour or morphology is demonstrated, an explanation is constructed to fit the facts, and subsequent citations quote the initial evidence as proof of the hypothesis that is derived from it.

Though many functional arguments are initially *post-hoc*, relatively few are *ad-hoc*: almost all claim generality of some kind and are consequently refutable by more detailed observation, by experiment or by examination of their validity in other groups of animals. While it is important that explanations should not be promoted from the status of speculation until such tests have been made, the fact that they originated in *post-hoc* explanations is not an objection: this is an inevitable stage in any observational science. In fact we rather doubt that *post-hoc* explanations are more misleading than many cases where predictions are formulated and subsequently tested: it is seldom difficult to produce facts that are generally congruent with a theory, and the process can appear falsely conclusive (see Northrop 1959).

The criteria for acceptance of adaptive explanations have already been widely described (Williams 1966; Curio 1973; Tinbergen 1963) and there is no need to review them. However, one of the most pervasive difficulties has received little detailed attention: confounding variables pose a fundamental problem to arguments that rely on comparative evidence. Their effects are frequently ignored (Gould 1978) and their scope and subtlety are often, we suspect, unappreciated.

Associations between ecological variables are common at all levels. For example, among rodents, intergeneric differences in comparative brain size are related to differences in habitat (forest-dwelling genera tend to have comparatively larger brains than grassland forms), diet (insectivores and frugivores have comparatively bigger brains than folivores), stratification (arboreal genera have comparatively bigger brains than terrestrial or fossorial ones) and activity timing (nocturnal genera tend to have comparatively bigger brains than diurnal ones) (Mace 1979). However, habitat, diet, stratification and activity timing are associated with each other. Though differences in each ecological variable are associated with differences in brain size when they are considered separately, when the effects of other variables are removed only the relation with diet remains.

In addition, morphological, physiological and behavioural characteristics also show associated distributions. A particular case where this is frequently forgotten concerns attempts to investigate the importance of kin selection in avian and mammalian social units. Such studies often seek to demonstrate that individuals treat relatives differently from other conspecifics (see, for example, Clutton-Brock & Harvey 1976; Kurland 1977; Greenwood & Harvey 1979). However, in many species, relatives are likely to spend much of their time close to each other as an indirect result of common social bonds to a joint matriarch: for example, among red deer, female sibs are often found in the same group because they retain close relationships with their mothers into adulthood (Clutton-Brock *et al.* 1979*c*). Proximity affects interaction frequency (see, for example, Clutton-Brock *et al.* 1976) and as a result, relatives may interact more often with each other than with unrelated animals. Consequently, evidence that they do so indicates that kin selection may have been involved only if differences in interaction rates exist when the proportion of time that relatives spend next to each other is taken into account (Clutton-Brock & Harvey 1976).

One of the firmest possible demonstrations that natural selection maintains the form or value of a trait is provided by evidence that natural variation between individuals is associated with differences in reproductive success or survival (Tinbergen 1965). For example, Kettlewell's demonstration that the melanic form of the peppered moth (*Biston betularia*) is less likely to be preyed upon by birds in areas of industrial pollution than the natural lighter-coloured morph while the situation is reversed in unpolluted areas, provides convincing evidence that the difference in coloration is an adaptation to avoiding predation by increasing crypticity (Kettlewell 1956, 1961). Long-term studies of populations of recognizable individuals make it possible to compare the reproductive success of individuals which differ in behaviour, physiology or morphology: over the last few years, studies have demonstrated associations between reproductive success or survival and differences in coloration (Curio 1973; Bantock *et al.* 1976), laying or birth date (Patterson 1965; Guinness *et al.* 1978*a*; Guinness *et al.* 1978*b*; Harvey *et al.* 1979), clutch size (Lack 1966), social dominance (Hausfater 1975; Packer 1977; Dunbar & Dunbar 1977), body size (Trivers 1972), mate fidelity (Coulson 1966), dispersal distance (Greenwood & Harvey 1976), territory size (Watson 1970; Watson & Miller 1971; Miller & Watson 1978) and horn size (Geist 1971).

While we believe that evidence of this kind is especially valuable in studying adaptation because it provides a detailed understanding of the mechanisms of natural selection, confounding variables can have widespread and complex effects. Two particular confounding variables commonly affect both reproductive success and many aspects of behaviour and morphology and can consequently produce non-causal associations between them: these are age and quality. The latter is difficult to define, though it is a term that would be recognized by any animal breeder: some individuals grow up small and sickly, they are prone to disease or behavioural aberrations and frequently develop small secondary sexual characteristics; others grow up large, strong and healthy and, in natural populations, tend to show superior reproductive success (see Coulson 1968; Geist 1971).

Let us start by examining a case where age appears to act as a confounding variable. In ungulates it is commonly claimed that differences in horn or antler size are related to fighting ability and reproductive success (see, for example, Beninde 1937; Geist 1971; Gould 1974; Maynard Smith, this symposium). For example, Geist (1966*b*, 1968, 1971) allocated bighorn rams (*Ovis canadensis*) to four categories based on horn size. His detailed observations showed that rams belonging to the category with largest horn size were most dominant and had the highest reproductive success. However, rams belonging to different categories varied in age and body size as well as in horn size (though the differences between the first and second highest categories were slight). Geist does not provide quantitative evidence that the association between horn size and reproductive success remained when differences in age and body size were taken into account and the possibility exists that the relation between them is a product of an association between horn size, body size and age. Among red deer (*Cervus elaphus*), reproductive success in rutting stags is correlated with antler length: the biggest harems are almost always held by mature stags with fully grown antlers (Darling 1937; Lincoln & Guinness 1973). However, antler length is related to age and body mass (Mitchell *et al.* 1977; Gibson 1978) and these are also related to reproductive success (Clutton-Brock *et al.* 1979*a*). It seems unlikely that individual differences in antler length are related to fighting ability: when antler length is correlated with dominance or reproductive success among stags of similar ages, no association is found (Clutton-Brock *et al.* 1979*a*).

The importance of taking age differences into account is sufficiently obvious to class as a truism, were it not so frequently ignored (Rowell 1974). For example, a wide variety of studies have shown that dominance in fighting success is correlated with reproductive success (Le Boeuf 1974; Packer 1977; Hausfater 1975; Dunbar & Dunbar 1977) but few have been able to control adequately for age effects. Similarly, several studies have shown that body size differences are correlated with dominance rank (see, for example, Tyler 1972). Both dominance and size are commonly age-related and, when the effects of age are removed, dominance rank is not always closely related to size (Clutton-Brock *et al.* 1976; Berger 1977).

Differences in quality between individuals can have equally far-reaching effects with the additional problem that, unlike age, they are often difficult to control for. For example, red deer calves associate more or less closely with their mothers during the first year of calf life (Guinness *et al.* 1979). About 10% of calves die in their first winter. Since mothers provide their offspring with food, shelter and protection from unrelated individuals and predators, we predicted that calves which spent much time with their mothers would have an increased chance of survival. Comparison of mortality among calves which associated more than average with their mothers with mortality of those that associated less than average (figure 1*a*) confirmed this prediction: calves that associated closely with their mothers were less likely to die.

However, examination of temporal changes in association convinced us that this was probably not a result of any benefits derived from maternal care. There was no significant relation between association frequency in the first months of calf life and subsequent mortality (figure 1*b*). In the months immediately before their death, deer calves that were going to die spent progressively less time with

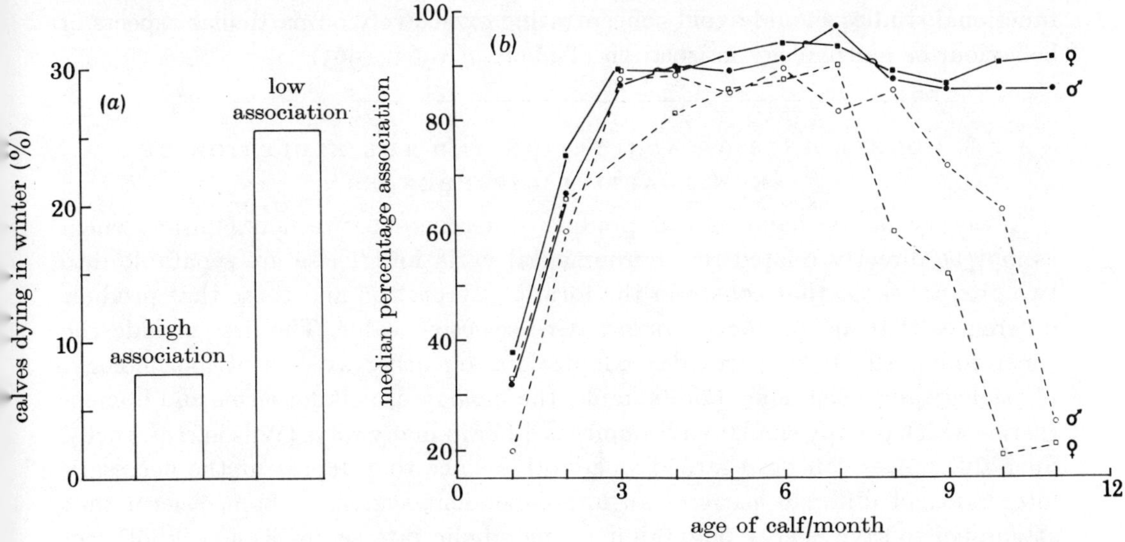

FIGURE 1. (*a*) Mortality among red deer calves which were recorded in the same feeding groups as their mothers more often than average for the population (high association) and less often (low association). (*b*) Median frequency of association with mother among red deer calves of different ages. The sample is divided into four groups: male calves that survived their first winter (●), male calves that died (○), females that survived (▪) and females that died (□).

their mothers. While absence from mother could have contributed to their increased chance of mortality, it seemed likely that the difference occurred because sickly calves tended to be left behind by their mothers and (because they were sickly) were also more likely to die. We were unable to control for 'sickliness' and so could not examine the effects of differences in association frequency.

Consistent inter-individual differences in resource access can also act as confounding variables. Among red deer hinds on Rhum, the availability of food in their winter home ranges varies widely. These differences are related to several other measures (including the proportion of daytime spent feeding, home range size, the average size of social groups, the birth weights and birth dates of calves and the frequency of mother–calf association) (Clutton-Brock *et al.* 1979*c*). In addition, they are related to fecundity and calf mortality. As a result, when we compare individuals, differences in each of the first five variables are related to reproductive success, though this association may not be the product of any causal relation and may occur purely because home range quality affects each of these variables as well as affecting reproductive success. Moreover, in red deer (and probably in many other mammals) quality differences can be transmitted across generations: not only do daughters commonly occupy the same home ranges of their mothers but poor mothers tend to produce poor offspring.

The problem of confounding variables is not one which has any easy solution. Confusion can only be minimized by care, intuition and an extensive knowledge of the kinds of variables that may be involved. It is consequently important that functional studies should avoid concentrating exclusively on particular aspects of behaviour or morphology in isolation (Tinbergen 1963, 1965).

3. Constraints on adaptation and the evolution of non-adaptive differences

A variety of mechanisms can produce differences between organisms which cannot be directly related to environmental variation. These are separable into two groups: those that constrain the form of adaptation and those that produce differences that are of neutral or negative selective value. The first include the constraining effects of particular adaptations on other aspects of morphology, physiology and behaviour. For example, the biology of poikilotherms and homeotherms which occupy similar environments differ in many ways (Wilson *et al.* 1978). Such differences can be regarded as adaptive since they represent the necessary integration of different parts of an interdependent system: a homeotherm that attempted to save energy maintaining a metabolic rate as low as a poikilotherm would be unlikely to reproduce successfully.

Mechanisms that can produce differences which have neutral or negative selective value include the results of genetic drift, linked or pleiotropic effects of selection operating on particular loci (including, presumably, heterogony: see below), delays in the process of adaptive change, the occupation of alternative adaptive peaks† (Lewontin 1978) and non-adaptive phenotypic modification (for example, the non-adaptive effects of unnaturally high population densities; see Christian 1963).

(*a*) *Constraints on adaptation*

Almost every adaptation constrains the size or form of some other aspect of the possessor's biology in some way (see, for example, McFarland 1976). Moreover, it is possible that particular combinations of traits may constrain other variables in ways that would be difficult to predict from the separate effects of their component parts.

The side-effects of particular adaptations are so widespread and so ramifying that they are frequently ignored or forgotten. Any attempts to review them would

† In fact, a case can be made that differences between 'alternative adaptive peaks' should mostly be regarded as adaptive on the grounds that species that take different adaptive pathways are likely to do so initially because some aspect of their biology affects selection pressures operating on subsequent changes. Consequently the two pathways do not represent true alternatives and differences between them should be regarded as the products of adaptive integration.

either be encyclopaedic or banal. Instead, we shall describe some of the repercussions of body size, a variable that has particularly far-reaching effects, bearing in mind that the size and form of many individual organs (or processes) may constrain other adaptations in related ways introducing similar complexities to comparative arguments.

Almost all anatomical measures are size-related, including the density and connectivity of cortical neurons (Jerison 1973), the presence of claws versus spatulate nails (see Napier & Napier 1967) and the degree of sexual dimorphism in body mass (Rensch 1959; Clutton-Brock *et al.* 1977). Many physiological variables, too, are size-related including, in particular, basal metabolic rate (Brody 1945), and most life history variables from gestation length to lifespan (Sacher 1959; Case 1978; Millar 1979; Western 1979). A wide range of behavioural variables are also size-related, including the rate of wing-flap in birds (Alexander 1971), the proportion of daytime spent in different activities (Clutton-Brock & Harvey 1977*a*), the rate of food intake (Sinclair 1977), diet quality (Bell 1971; Jarman 1974; Janis 1976) and home range size (McNab 1963; Schoener 1968; Clutton-Brock & Harvey 1977*b*). We predict that many other variables will also prove to be related to size including rates of interaction, the duration and frequency of bouts of suckling and copulation, the probability of damage in fights, the upper limits of the rate of vocalization and the lower limits of pitch as well as most aspects of behavioural development.

Such relations pose a problem to comparative studies. While it is important that we should ask functional questions about size itself (Bourlière 1975), if we wish to examine relations between traits that are inevitably affected by body size and ecology or social behaviour, we need to remove the effects of size first. This cannot usually be achieved by dividing the value of the variable by body size (either length or, if it is a cubic measure, by body mass or volume) since most biological parameters do not increase in a 1:1 relation with body size: some increase faster while others increase more slowly (Huxley 1932; Goss 1964; Alexander 1971). For example, the brain mass of different bird species increases at around the 0.60 power of body mass, the mass of their flight muscles at around the 0.96 power while the rate at which they flap their wings decreases at around the -0.26 power (Alexander 1971). Variables of this kind whose values depend on overall body size are referred to as allometric (see Gould 1966; see also note added in proof).

The problem that allometry raises is not insuperable. It is possible to remove allometric effects by calculating the appropriate regression (sometimes with transformed variables) of any variable on body size and to measure deviations from this regression line: relations between these deviations and ecological variables can then be examined (see, for example, Pirlot & Stephan 1970; Eisenberg & Wilson 1979).

However, for deviation measurements to be biologically meaningful, we need to solve two further problems. First, we must distinguish between variables that are inevitably dependent on body size and those that are merely correlated with it for independent adaptive reasons. In the first case, we shall need to remove the effects of size by using deviation measurements of some kind, while in the second this approach would be misleading. For example, it is necessary for larger animals to have relatively stronger, heavier skeletons than smaller ones (Schmidt-Nielsen 1975) and if we wished to examine the ecological correlates of skeletal mass, we should remove the effects of size first. In several mammalian Orders, interspecific differences in the size of social groupings are also positively correlated with body size (Jarman 1974; Clutton-Brock & Harvey 1977*b*). While the reasons for this association are not fully understood, it seems likely that large body size and large group size represent independent adaptations to habitats where food is widely dispersed or predators are common (Jarman 1974; Geist 1974*a*, *b*). In this case, if we wished to examine the ecological correlates of group size, it would be misleading to use deviations from the 'allometric' slope of group size on body size since there is no reason to suppose that body size acts as a primary determinant of group size.

There are no firm criteria for distinguishing between relations of these two types. An understanding of the physical effects of size is obviously important. In addition, a knowledge of the nature of the relation in several animal groups as well as of the amount of variance that is accounted for by body size can also be indicative: in general, variables that are size-dependent show very close relationships with size while those that are correlated with it for independent reasons show weaker correlations (see Clutton-Brock & Harvey 1977*a*, *b*; Clutton-Brock & Harvey 1979).

The second main problem of comparing size-dependent variables between species centres on the fact that allometric relations differ between phylogenetic groups (see Gould 1975; Radinsky 1978). When many biological variables are regressed on size in different groups of organisms, marked differences in elevation (and, less commonly, differences in slope) are frequently found between groups. For example, families of birds differ consistently in the relation between body size and relative egg mass (Lack 1968). Similar differences exist among mammals: if gestation length, litter mass, weaning age, at sexual maturity, lifespan or brain size are regressed on body mass in different families of primates or rodents, marked differences in elevation are found between families (Clutton-Brock & Harvey 1979; Clutton-Brock *et al.* 1979*d*; Mace 1979).

Moreover, the slopes of relations calculated across species within families tend to be less steep than those calculated across phylogenetically heterogeneous assemblages. Consequently, smaller species within any family tend to show positive deviations from the overall regression line (see figure 2) and large species tend to show negative deviations. As a result, such deviation measures across families are not independent of size within families, and are likely to be correlated to ecological variables related to size.

The problem extends beyond inter-family comparisons. It appears to be a common biological phenomenon that size-dependent variables show progressively shallower slopes, when regressed on body size, the lower the taxonomic level that is considered. For example, if brain mass is regressed on body mass across individuals of the same species, slopes are shallower than if regressions are calculated across mean values for different congeneric species – and regressions calculated across these, in turn, produce shallower slopes than those calculated from regressions across all species belonging to the same family (see Gould (1975) for a review of the topic). Consequently, even if the problem of inter-family differences is solved by calculating, for each species, deviations from *the regression line calculated for its family alone*, large species belonging to the same genus will tend to show lower deviations compared with small ones.

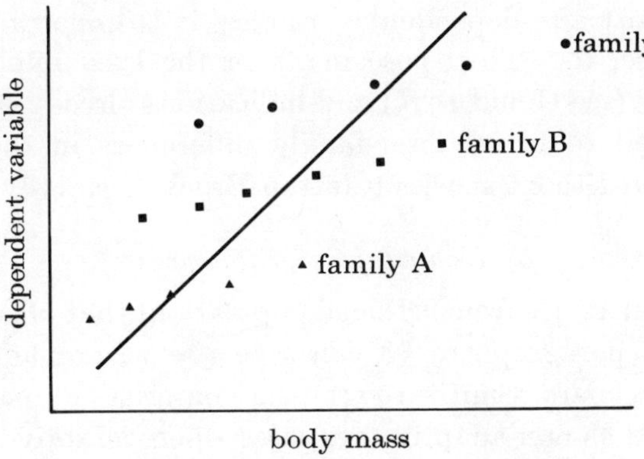

FIGURE 2. Idealized relation between logarithm of a size-dependant variable and logarithm of body size in four families of a hypothetical order. Though differences in elevation and slope have been exaggerated, the figure provides a realistic indication of the relations between size and most life history variables both in primates and rodents (see Clutton-Brock *et al.* 1979*d*; Mace 1979).

One solution to this problem is to restrict comparisons to the taxonomic level immediately below that at which the regression line has been set (see Clutton-Brock & Harvey 1979). For example, if it is necessary to compare congeneric species, a regression line can be set for the genus by using the mean values for each species. Deviations from this line for each species can be compared, but deviations from the same line calculated for different individuals, populations or subspecies will still be size-related. Similarly, if genera are to be compared, lines can be set by using mean values for different genera belonging to the same family. Deviations can then be compared between genera, but those calculated for congeneric species or subspecies will still be affected by size. Though it is possible to determine the level at which regression lines can most usefully be set by analysis of variance nested at different taxonomic levels (see Clutton-Brock & Harvey 1977*b*), in most cases the level of comparison is determined by the amount of data available or by the aims of the study.

Having removed the effects of size in this way, it is possible to demonstrate relations between relative values of size-dependent variables and ecological differences that are not intuitively obvious. In some cases, similar relations appear to exist in phylogenetically diverse orders. For example, in primates and rodents respectively, folivores and grazers tend to show comparatively small brain sizes for their body masses (Clutton-Brock & Harvey 1979; Mace 1979). Comparative values of life history variables are also related to diet: among rodents, grazing genera have comparatively short gestation lengths, earlier weaning ages, earlier sexual maturity and shorter lifespans than other species (Mace 1979) and in primates, too, there is some evidence of similar trends among folivores (Clutton-Brock *et al.* 1979*d*).

Since inter-family differences in elevation account for a large proportion of the total variance in many size-dependent variables, it is important that we should seek to explain these, too. They pose many of the least explored questions in evolutionary ecology (see Gould 1975) and indications already exist that some, at least, may be related to broad inter-family differences in the types of niches occupied by their constituent species (Clutton-Brock & Harvey 1979).

(*b*) *Non-adaptive differences*

Although there is evidence from artificial populations that all six of the processes listed in the second paragraph of §3 can give rise to non-adaptive differences (Lewontin 1970; Maynard Smith 1975), the majority of naturally occurring differences postulated as non-adaptive are cases where adaptive arguments fail to account for the observed situation. For example, among closely related species of the Callitrichidae (marmosets and tamarins) which occupy different parts of the Amazon basin, complex differences in coloration exist that are apparently unrelated to variation in climate, habitat, diet or social system (Hershkowitz 1977): it seems likely that they may be the product either of drift or of slight differences in coloration combined with selection for mechanisms that isolate species. At a higher taxonomic level, the Bovidae have developed horns formed from specialized ossicones as well as from skin outgrowths. In contrast, the antlers of the Cervidae, which serve similar functions, have developed from an outgrowth of the frontal bone (Geist 1966*a*). There is little evidence that this difference is attributable to ecological differences between the two families, and horns and antlers may represent alternative solutions to the same problem. In the feral pigeon, *Columba livia* (Murton *et al.* 1973; Murton *et al.* 1974), dark coloured forms are common and are spreading in certain areas. Dark coloration is associated with improved reproductive success, partly because dark morphs can apparently breed earlier than pale coloured forms. Unlike the peppered moth (see above), there is no indication that this is the direct result of any beneficial effect of dark coloration and it may occur because a gene (or genes) producing dark coloration is linked with a gene (or genes) which affects the date of breeding. Murton speculates that a similar linkage group might explain the difference between the light-coloured peregrine falcon (*Falco peregrinus*) and its dark-coloured congenerics the sooty falcon and Eleonora's falcon (*F. concolor* and *F. eleonorae*), which, unlike the peregrine, breed in early autumn, capitalizing on the supply of passage migrants then available (Murton *et al.* 1974). Comparisons of interspecific differences sometimes suggest cases where a species or group of species may have failed, so far, to produce an adaptation present in related taxa. Among the marmosets and tamarins, Goeldi's monkey (*Callimico goeldii*) produces a single offspring whereas all other species usually produce twins (Hershkowitz 1977). One might predict that *Callimico* would have compensated for this by an increase in (relative) neonatal mass, a reduction in gestation length or a reduction in relative weaning age. However, it shows none of these, perhaps because changes in the fundamental growth process require major ontogenetic reorganization.

All four of these non-adaptive explanations are tenuous and, in each case, it is possible that the difference may eventually prove to be adaptive as many examples of 'non-adaptive' differences have done in the past. Because the identification of non-adaptive differences relies on the absence of evidence for adaptation, it is unfortunate that many postulated non-adaptive differences involve traits whose functions are even less well known. For example, in a recent review of the problems of adaptation, Lewontin (1978) selects the difference in horn number between the African and the Indian rhinoceros as a case of alternative adaptive peaks. 'For example', he argues, 'the Indian rhinoceros has one horn and the African rhinoceros has two. Horns are an adaptation for protection against predators, but it is not true that one horn is specifically adaptive under Indian conditions as opposed to two horns on the African plains.' However, no one has yet compared predation pressures on African and Indian rhinos, estimated the efficacy of horns in defence or measured the costs of growing them in plains country versus forest. Indeed, it seems likely that they may represent an adaptation to intra-specific combat rather than predator pressure (see Ullrich 1964; Owen-Smith 1977; Goddard 1967). Any biologist familiar with the ways in which apparently irrelevant details of behaviour and morphology can prove, on closer examination, to have important functions is likely to agree with Tinbergen's conclusion that field studies 'show how useless and even unscientific it is to pronounce an opinion on the question whether or not "a certain feature is adaptive" until one has tried to find out' (Tinbergen 1965).

There is little point in reviewing here the variety of non-adaptive differences which detailed research has subsequently shown to be adaptive (see Cain 1964). However, it is worth examining one case, because it is related to the effects of body size and is commonly confused with allometry. In the early years of the twentieth century, several biologists noted that certain secondary sexual characteristics, including horns and antlers, tended to be larger, relative to body size, in larger species (Lameere 1904, 1915; Smith 1906). Research also showed that during the process of individual ontogeny some of these organs grew faster than overall body size, so that if measurements were taken at different age stages, positive allometry was evident: such organs were called 'heterogonic' (Pézard 1921). Both sets of observations were brought together by Huxley (1932): selection for increased body size, he argued, automatically increases the relative size of heterogonic organs and, as a result, they are larger in larger species. His theory was supported by the tendency for heterogonic organs to show positive allometry when plotted across mature individuals of a given species (see, for example, Huxley 1931). As Huxley realized, trends of this kind could, if general, provide a major source of non-adaptive differences: 'Whenever we find the rule of Lameere and Smith [that secondary sexual characteristics increase in size in larger species] holding true, we are justified in concluding that the relative size of horn, mandible, or other heterogonic organ is automatically determined as a secondary result of a common growth-mechanism and *therefore is not of adaptive significance*. This provides us with a large new list of non-adaptive specific and generic characters' (original italics).

One of the classic cases of heterogony (Huxley 1932) concerns antler size in deer: antler size shows a positive allometric relation to body size both during the process of individual ontogeny, across mature individuals of the same species (though not in all cases) and across different species belonging to the same family (Cervidae). However, investigation of the relationship between relative antler size and breeding system suggests that the interspecific trend has an adaptive basis and is unlikely to be an inevitable side-effect of selection for increasing body size (Clutton-Brock *et al.* 1979*b*): large deer species tend to be more strongly polygynous than small ones, which are mostly semi-monogamous. Males of polygynous species probably compete more intensively for breeding access and may, as a result, invest in relatively large antlers. Though it seems unlikely that similar heterogonic relations across invertebrate species can be explained on identical grounds, we speculate that larger species (which typically have longer lifespans than smaller ones) may fight more frequently during their lives and, for this reason, may also be prepared to invest relatively more in aggressive and defensive apparatus (Clutton-Brock *et al.* 1979*b*). Whether all cases of interspecific heterogony will eventually prove to be adaptive remains to be seen.

One way of distinguishing between adaptive and non-adaptive variation is to examine relations between morphological, physiological or behavioural variables and ecological variables across groups of species (Lack 1968). If the distribution of a trait can be predicted on ecological criteria, then there are reasonable grounds for supposing that at least a part of the variation is adaptive. For example, recent studies of primates have shown that species differences in home range size are closely related to energetic requirements of the group (Clutton-Brock & Harvey 1977*a*), that polygynous species tend to show greater sexual dimorphism in body size (Clutton-Brock *et al.* 1977) and relative canine size (Harvey *et al.* 1978*a*, *b*) and that folivorous species tend to have relatively small brains compared to frugivores (Clutton-Brock & Harvey 1979).

Though they face a variety of methodological problems (Clutton-Brock & Harvey 1977*a*, *b*) investigations of broad adaptive trends have other advantages.

It is possible to account for the effects of body size as well as for phylogenetic differences in allometric relations. Associations between ecological variables can be investigated and, in some cases, the effects of associated variables can be distinguished (Clutton-Brock & Harvey 1977*a*). And, as more species are considered, it becomes progressively more difficult to fit several feasible adaptive hypotheses to the empirical facts. For example, two recent studies of particular rodent species have theorized about the reproductive strategies of desert-living rodents. The first (Schwartz & Bleich 1975) argued that harsh environments select for quick breeding and fast growth rates. The second (Layne 1966), suggested that they select for slow breeding and slow growth rates. While it is entirely possible that harsh environments may select for both strategies, a quantitative comparison of desert-living rodent species with woodland or forest forms would have determined which generalization, if either, was correct and which strategy represented an exception to the general trend. Finally, an understanding of basic adaptive trends often indicates which selective pressures are likely to have affected particular kinds of variables. Only when we understand the ground rules of adaptation can we hope to explain both the form of general trends and the exceptions to them.

We have discussed some of the difficulties that face adaptive explanations based on comparative evidence. There are many others. As Gould (1978) has clearly pointed out, it is important that evolutionary biologists should seek to solve these problems if adaptive arguments are to retain credibility. No better object lesson is needed than Tinbergen's (1964) description of his undergraduate days: 'In the post-Darwinian era, a reaction against uncritical acceptance of the selection theory set in, which reached its climax in the great days of Comparative Anatomy, but which still affects many physiologically inclined biologists. It was a reaction against making uncritical guesses about the survival value, the function, of life processes and structures. This reaction, of course healthy in itself, did not (as one might expect) result in an attempt to improve methods of studying survival value; rather it deteriorated into lack of interest in the problem – one of the most depolarable things that can happen to a science.'

Note added in proof, 21 *June* 1979 (see §3*a*). Gould refers to all size-related variables as allometric, whether they are dependent on size or correlated with size for independent, adaptive reasons. In this review, we use the term to refer to the former category only.

We should like to thank Mr S. D. Albon, Dr P. P. G. Bateson, Dr B. C. R. Bertram, Dr R. I. M. Dunbar, Professor S. J. Gould, Mr J. Hartung, Dr P. J. Jarman, Dr N. Knowlton, Miss G. M. Mace, Dr G. A. Parker and Professor J. Maynard Smith, F.R.S., for help, advice or criticism. We are particularly grateful to Mr S. D. Albon, Miss F. E. Guinness and Miss G. M. Mace for permission to quote unpublished results.

References (Clutton-Brock & Harvey)

Alexander, R. M. 1971 *Size and shape*. London: Edward Arnold.

Altmann, S. A. 1971 *Social communication in primates*. University of Chicago Press.

Bantock, C. R., Bayley, J. A. & Harvey, P. H. 1976 Simultaneous selective predation on two features of a mixed sibling species population. *Evolution* **29**, 636–649.

Bell, R. H. V. 1971 A grazing system in the Serengeti. *Scient. Am.* **225**, 86–93.

Beninde, J. 1937 Naturgeschichte des Rothirsches. *Monographie Wildsäugetiere*, vol. 4. Leipzig: P. Schops.

Berger, J. 1977 Organisational systems and dominance in feral horses in the Grand Canyon. *Behav. Ecol. Sociobiol.* **2**, 131–146.

Bourlière, F. 1975 Mammals, small and large: the ecological implications of size. In *Small mammals, their productivity and population dynamics* (ed. F. B. Golley, K. Petrusiwicz & L. Ryszkowski), pp. 1–8. Cambridge University Press.

Braestrup, F. W. 1971 The evolutionary significance of learning. *Vidensk. Meddr dansk naturh. Foren.* **134**, 89–102.

Brody, S. 1945 *Bioenergetics and growth*. New York: Reinhold.

Cain, A. J. 1964 The perfection of animals. *Viewpoints Biol.* **3**, 37–63.

Carpenter, C. R. 1973 *Behavioural regulators in behaviour of primates*. Lewisburg: Bucknell University Press.

Case, T. J. 1978 On the evolution and adaptive significance of postnatal growth rates in the terrestrial vertebrates. *Q. Rev. Biol.* **53**, 243–282.

Christian, J. J. 1963 The pathology of overpopulation. *Milit. Med.* **128**, 571–603.

Clutton-Brock, T. H. 1977 Some aspects of intraspecific variation in feeding and ranging behaviour in primates. In *Primate ecology: studies of feeding and ranging behaviour in lemurs, monkeys and apes* (ed. T. H. Clutton-Brock), pp. 539–556. London: Academic Press.

Clutton-Brock, T. H. & Harvey, P. H. 1976 Evolutionary rules and primate societies. In *Growing points in ethology* (ed. P. P. G. Bateson & R. A. Hinde), pp. 195–237. Cambridge University Press.

Clutton-Brock, T. H. & Harvey, P. H. 1977*a* Species differences in feeding and ranging behaviour in primates. In *Primate ecology: studies of feeding and ranging behaviour in lemurs, monkeys and apes* (ed. T. H. Clutton-Brock), pp. 556–584. London: Academic Press.

Clutton-Brock, T. H. & Harvey, P. H. 1977*b* Primate ecology and social organisation. *J. Zool.* **183**, 1–39.

Clutton-Brock, T. H. & Harvey, P. H. 1979 Primates, brains and ecology. *J. Zool.* (In the press.)

Clutton-Brock, T. H., Albon, S. D., Gibson, R. M. & Guinness, F. E. 1979*a* The logical stag: adaptive aspects of fighting in red deer (*Cervus elaphus* L.). *Anim. Behav.* **27**, 211–225.

Clutton-Brock, T. H., Albon, S. D. & Harvey, P. H. 1979*b* Antlers, body size and breeding systems in the cervidae. In preparation.

Clutton-Brock, T. H., Greenwood, P. J. & Powell, R. P. 1976 Ranks and relationships in highland ponies and highland cows. *Z. Tierpsychol.* **41**, 202–216.

Clutton-Brock, T. H., Guinness, F. E. & Albon, S. D. 1979*c* The red deer of Rhum. In preparation.

Clutton-Brock, T. H., Harvey, P. H. & Rudder, B. 1977 Sexual dimorphism, socionomic sex ratio and body weight in primates. *Nature, Lond.* **269**, 797–799.

Clutton-Brock, T. H., Harvey, P. H. & Rudder, B. 1979*d* Adaptive aspects of life history variables in primates. In preparation.

Coulson, J. C. 1966 The influence of the pair bond and age on the breeding biology of the kittiwake gull *Rissa tridactyla*. *Ecology* **47**, 269–279.

Coulson, J. C. 1968 Differences in the quality of birds nesting in the centre and on the edge of a colony. *Nature, Lond.* **217**, 478–499.

Curio, E. B. 1973 Towards a methodology of teleonomy. *Experientia* **29**, 1045–1058.

Darling, F. F. 1937 *A herd of red deer*. Oxford University Press.
Darwin, C. 1859 *On the origin of species*. London: John Murray.
Darwin, C. 1871 *The descent of man and selection in relation to sex*. London: John Murray.
Dunbar, R. I. M. & Dunbar, P. 1977 Dominance and reproductive success among female gelada baboons. *Nature, Lond.* **266**, 351–352.
Eisenberg, J. F. & Wilson, D. E. 1978 Relative brain size and feeding strategies in the Chiroptera. *Evolution* **32**, 740–751.
Geist, V. 1966*a* The evolution of horn-like organs. *Behaviour* **27**, 175–214.
Geist, V. 1966*b* The evolutionary significance of mountain sheep horns. *Evolution* **20**, 558–566.
Geist, V. 1968 On the interrelation of external appearance, social behaviour and social structure of mountain sheep. *Z. Tierpsychol.* **25**, 199–215.
Geist, V. 1971 *Mountain sheep*. University of Chicago Press.
Geist, V. 1974*a* On the relationship of social evolution and ecology in ungulates. *Am. Zool.* **14**, 205–220.
Geist, V. 1974*b* On the relationship of ecology and behaviour in the evolution of ungulates: theoretical consideration. In *The behaviour of ungulates and its relation to their management* (ed. V. Geist & F. Walther), pp. 235–246. Morges: I.U.C.N.
Gibson, R. M. 1978 Behavioural factors affecting reproductive success in red deer stags. D.Phil. thesis, University of Sussex.
Goddard, J. 1967 Home range, behaviour and recruitment rates of two black rhinoceros populations. *E. Afr. Wildl. J.* **5**, 133–150.
Goss, R. J. 1964 *Adaptive growth*. London: Logos Press.
Gould, S. J. 1966 Allometry and size in ontogeny and phylogeny. *Biol. Rev.* **41**, 587–640.
Gould, S. J. 1974 The evolutionary significance of 'bizarre' structures: antler size and skull size in the 'Irish Elk', *Megaloceros giganteus*. *Evolution* **28**, 191–220.
Gould, S. J. 1975 Allometry in primates with emphasis on scaling and the evolution of the brain. *Contr. Primatol.* **5**, 244–292.
Gould, S. J. 1978 Sociobiology: the art of story telling. *New Scient.* **80**, 530–533.
Greenwood, P. J. & Harvey, P. H. 1976 Differential mortality and dispersal of male blackbirds. *Ringing & Migration* **1**, 75–77.
Greenwood, P. J. & Harvey, P. H. 1979 Dispersal and philopatry: an evolutionary compromise? In preparation.
Guiness, F. E., Albon, S. D. & Clutton-Brock, T. H. 1978*a* Factors affecting reproduction in red deer (*Cervus elaphus* L.). *J. Reprod. Fert.* **54**, 325–334.
Guinness, F. E., Clutton-Brock, T. H. & Albon, S. D. 1978*b* Factors affecting calf mortality in red deer. *J. Anim. Ecol.* (In the press.)
Guinness, F. E., Hall, M. J. & Cockerill, R. C. 1979 Mother–offspring association in red deer. *Anim. Behav.* (In the press.)
Hamilton, W. D. 1964*a* The genetical evolution of social behaviour I. *J. theor. Biol.* **7**, 1–16.
Hamilton, W. D. 1964*b* The general evolution of social behaviour II. *J. theor. Biol.* **7**, 17–52.
Harvey, P. H., Greenwood, P. J., Perrins, C. M. & Martin, A. R. 1979 Breeding success of great tits in relation to age of male and female parent. *Ibis* **121**, 216–219.
Harvey, P. H., Kavanagh, M. & Clutton-Brock, T. H. 1978*a* Sexual dimorphism in primate teeth. *J. Zool.* **186**, 475–486.
Harvey, P. H., Kavanagh, M. & Clutton-Brock, T. H. 1978*b* Canine tooth size in female primates. *Nature, Lond.* **276**, 817–818.
Hausfater, G. 1975 Dominance and reproduction in baboons (*Papio cynocephalus*): a quantitative analysis. *Contr. Primatol.* **7**, 1–150.
Hershkovitz, P. 1977 *Living new world monkeys*, vol. 1: *Platyrrhini*. University of Chicago Press.
Hinde, B. A. 1970 *Animal behaviour*. New York: McGraw-Hill.
Hinde, B. A. 1975 The concept of function. In *Function and evolution of behaviour* (ed. G. Baerends, C. Beer & A. Manning), pp. 3–15. Oxford: Clarendon Press.
Huxley, J. S. 1931 The relative size of antlers in deer. *Proc. zool. Soc. Lond.* **19**, 819–864.

Huxley, J. S. 1932 *Problems of relative growth*. London: MacVeagh.
Janis, C. 1976 The evolutionary strategy of the Equidae and the origins of rumen and caecal digestion. *Evolution* **30**, 757–774.
Jarman, P. J. 1974 The social organisation of antelope in relation to their ecology. *Behaviour* **48**, 215–267.
Jerison, H. J. 1973 *Evolution of the brain and intelligence*. New York: Academic Press.
Kettlewell, H. B. D. 1956 Further selection experiments on industrial melanism in the Lepidoptera. *Heredity, Lond.* **10**, 287–301.
Kettlewell, H. B. D. 1961 The phenomenon of industrial melanism in Lepidoptera. *A. Rev. Ent.* **6**, 245–262.
Kummer, H. 1971 *Primate societies: group techniques of ecological adaptation*. Chicago: Aldine Atherton.
Kurland, J. A. 1977 Kin selection in the Japanese monkey. *Contr. Primatol.* **12**, 1–145.
Lack, D. 1966 *Population studies of birds*. Oxford University Press.
Lack, D. 1968 *Ecological adaptations for breeding in birds*. London: Methuen.
Lameere, A. 1904 L'évolution des ornaments sexuels. *Bull. Acad. r. Belg. Cl. Sci.*, p. 1337.
Lameere, A. 1915 Les caractères sexuels des Prionides. *Bull. scient. Fr. Belg.* **49**, 1915.
Layne, J. N. 1966 Postnatal growth and development of *Peromyscus floridanus*. *Growth* **30**, 23–45.
LeBoeuf, B. J. 1974 Male–male competition and reproductive success in elephant seals. *Am. Zool.* **14**, 163–176.
Lewontin, R. C. 1970 *The genetic basis of evolutionary change*. New York: Columbia University Press.
Lewontin, R. C. 1978 Adaptation. *Scient. Am.* **239** (3), 156–169.
Lewontin, R. C. 1979 Sociobiology as an adaptationist program. *Behav. Sci.* **24**, 1–10.
Lincoln, G. A. & Guinness, F. E. 1973 The sexual significance of the rut in red deer. *J. Reprod. Fert. Suppl.* **19**, 475–489.
Mace, G. M. 1979 The evolutionary ecology of small mammals. D.Phil. thesis, University of Sussex.
McFarland, D. 1976 Form and function in the temporal organisation of behaviour. In *Growing points in ethology* (ed. P. P. G. Bateson & R. A. Hinde), pp. 55–93. Cambridge University Press.
McNab, B. N. 1963 Bioenergetics and the determination of home range size. *Am. Nat.* **97**, 133–140.
Maynard Smith, J. 1975 *The theory of evolution*. Harmondsworth: Penguin Books.
Millar, J. S. 1979 Body size and reproduction in terrestrial eutherian mammals. Evolution. (In the press.)
Miller, G. R. & Watson, A. 1978 Territories and the food plants of individual red grouse. I. Territory size, number of males, and brood size compared with the abundance, production and diversity of heather. *J. Anim. Ecol.* **47**, 293–307.
Mitchell, B. M., Staines, B. W. & Welch, D. 1977 *Ecology of red deer: a research review relevant to their management in Scotland*. Cambridge: Institute of Terrestrial Ecology.
Murton, R. K., Thearle, R. J. P. & Coombs, C. F. B. 1974 Ecological studies of the feral pigeon *Columba livia* var. III. Reproduction and plumage polymorphism. *J. appl. Ecol.* **11**, 841–852.
Murton, R. K., Westwood, N. J. & Thearle, R. J. P. 1973 Polymorphism and the evolution of continuous breeding seasons in the pigeon *Columba livia*. *J. Reprod. Fert. Suppl.* **19**, 561–575.
Napier, J. R. & Napier, P. H. 1967 *A handbook of living primates*. London: Academic Press.
Northrop, F. S. C. 1959 *The logic of the sciences and of the humanities*. New York: Meridian Books.
Owen-Smith, N. 1977 On territoriality in ungulates and an evolutionary model. *Q. Rev. Biol.* **52**, 1–38.
Packer, C. R. 1977 Inter-troop transfer and inbreeding avoidance in *Papio anubis* in Tanzania. Ph.D. thesis, University of Sussex.

Patterson, I. J. 1965 Timing and spacing of broods in the black-headed gull, *Larus ridibundus*. *Ibis* **107**, 433–459.

Pézard, A. 1921 Numerical law of regression of certain secondary sex characters. *J. gen. Physiol.* **3**, 271.

Pirlot, P. & Stephan, H. 1970 Encephalization in Chiroptera. *Can. J. Zool.* **48**, 433–438.

Radinsky, L. 1978 Evolution of brain size in carnivores and ungulates. *Am. Nat.* **112**, 814–831.

Rensch, B. 1959 *Evolution above the species level*. London: Methuen.

Rowell, T. E. 1974 The concept of social dominance. *Behav. Biol.* **11**, 131–154.

Sacher, G. A. 1959 Relationship of lifespan to brain weight and body weight in mammals. In *C.I.B.A. Foundation Symposium on the Lifespans of Animals* (ed. G. E. W. Wolstenholme & M. O'Connor), pp. 115–133. Boston, Mass.: Little, Brown and Co.

Schmidt-Nielsen, K. 1975 Scaling in biology: the consequences of size. *J. exp. Zool.* **194**, 287–307.

Schoener, T. W. 1968 Sizes of feeding territories among birds. *Ecology* **49**, 123–141.

Schwartz, O. A. & Bleich, V. C. 1975 Comparative growth in two species of woodrats, *Neotoma lepida intermedia* and *Neotoma albigula venusta*. *J. Mammal.* **56**, 653–666.

Sinclair, A. R. E. 1977 *The African buffalo: a study of resource limitation of populations.* University of Chicago Press.

Smith, G. W. 1906 High and low dimorphism. *Mitt. zool. Stn Neapel* **17**, 312.

Tinbergen, N. 1963 On aims and methods of ethology. *Z. Tierpsychol.* **20**, 410–433.

Tinbergen, N. 1965 Behaviour and natural selection. In *Ideas in modern biology* (ed. J. A. Moore), pp. 521–542. New York: Natural History Press.

Tinbergen, N. 1972 Functional ethology and the human sciences. *Proc. R. Soc. Lond.* B **182**, 385–410.

Trivers, R. L. 1972 Parental investment and sexual selection. In *Sexual selection and the descent of man, 1871–1971* (ed. B. G. Campbell), pp. 136–179. Chicago: Aldine-Atherton.

Tyler, S. 1972 The social organisation of New Forest ponies. *Anim. Behav. Monogr.* **5**, 87–196.

Ullrich, W. 1964 Zur biologie der Panzernashörner (*Rhinoceros unicornis*) in Assam. *Zool. Gart., Lpz* **28**, 225–250.

Watson, A. 1970 Territorial and reproductive behaviour of red grouse. *J. Reprod. Fert. Suppl.* **11**, 3–14.

Watson, A. & Miller, G. R. 1971 Territory size and aggression in a fluctuating red grouse population. *J. Anim. Ecol.* **40**, 367–383.

Western, D. 1979 Life history and ecological implications of size in mammals. In preparation.

Williams, G. C. 1966 *Adaptation and natural selection, a critique of some current evolutionary thought*. Princeton University Press.

Wilson, E. O. (ed.) 1978 *Life on Earth*, 2nd edn. Sunderland, Mass.: Sinauer Associates

Proc. R. Soc. Lond. B **205**, 567–580 (1979)
Printed in Great Britain

The question of adaptive sex ratio in outcrossed vertebrates

BY G. C. WILLIAMS

Department of Ecology and Evolution, State University of New York, Stony Brook, New York 11794, *U.S.A.*

Of various published theories of adaptive control of progeny sex ratio only two are plausible, a physiological theory by Trivers & Willard, and a demographic theory by Verner. The first applies to species in which sons and daughters impose different costs on parents, and in which only one or very few young are produced at once. They ought to show positive correlations in the sex of successive offspring and high sex-ratio variance among progenies. Verner's theory postulates a minimization of competition for mates in neighbourhoods subject to random fluctuation in sex ratio. Optimal progenies would exactly match the population's evolutionary equilibrium sex ratio. There would be little variance among progenies.

Evidence from vertebrates is unfavourable to either theory and supports, instead, a non-adaptive model, the purely random (Mendelian) determination of sex. The apparent absence of parental control of progeny sex ratio is a serious theoretical difficulty.

It is often assumed that offspring sex ratio is important for parental fitness and that the best parental strategy is to devote resources equally to the rearing of sons and daughters. This paper discusses the reasoning behind this and other theories of adaptive sex ratio and reviews some data that seem to challenge their validity. Both reasoning and data relate explicitly to outcrossed vertebrates, but are applicable to other organisms with similar life cycles. Unless otherwise stated I assume an X–Y sex determining mechanism, autosomal influences on offspring sex ratio, large outcrossed populations, and sons and daughters that are equally costly at all stages of parental investment.

Under these conditions, selection will result in equal numbers of males and females up to the end of parental investment. More exactly, the parents collectively are expected to invest equally in sons and daughters. This conclusion was reached first by Fisher (1930) and then refined and extended by many others (e.g. Shaw 1958; MacArthur 1965; Charnov 1975). The arguments vary in detail, but all are based on the same essential premise of a mating advantage for the minority sex. If, for instance, males are scarce in any population, they will on average have more offspring than females. Thus any genetic tendency to be a male, or to produce sons rather than daughters, will be favoured by selection. So males will increase until they are no longer a minority. Neither sex will have an advantage in a population at its equilibrium of equal investment in sons and daughters. Usually this will mean very nearly equal numbers of males and females at most stages of development. The robustness and broad applicability of this conclusion, and its insensitivity to moderate alterations in its premises were recently emphasized by Eshel (1975).

Careful reading of the original arguments shows (with a few exceptions reviewed below) that they justify only this conclusion on total son and daughter investments by the parents collectively. They are silent on matters of parental strategy within a population at equilibrium. Processes discussed by Fisher, Shaw, and others can be regarded as frequency-dependent selection with offspring sex ratio as a neutral character at evolutionary equilibrium. Note that parents collectively could produce equal numbers of boys and girls by having some couples with nothing but sons and others nothing but daughters, by having every couple produce exactly the same number of sons and daughters, and by many other arrangements. The inapplicability of a population equilibrium argument to individual fitness, and the distinction between population values produced by selection, and characters favoured by selection within populations, are often overlooked by some distinguished theorists. For instance, Wilson (1975, p. 317) clearly summarized the reasons why a population should be equally productive of males and females, and then immediately discussed these equal numbers as the 'optimum sex ratio' for a parent.

My usage of such concepts as optimality and adaptation will be a bit more restrictive than is found in other treatments of these and related concepts. For instance, an evolutionarily stable strategy is sometimes defined as a feature that is invulnerable to any degree of replacement by an alternative feature of greater fitness. This would seem to include what might be termed evolutionarily neutral strategies. In this paper a neutral equilibrium would not be considered adaptive. Adaptation will imply a finite superiority to alternatives and a consequent invulnerability to degeneration by random processes. The optimum sex ratio for a parent's progeny would be the one likely to result in the maximum number of grandchildren. In a population with an excess of females, for example, the optimum progeny would be 100% male.

The naïve Mendelian theory

Discussions in textbooks often imply that progeny sex ratio is merely a mechanical consequence of cytological machinery and without adaptive significance to the parents. A cross between a male and a female is seen as a Mendelian test cross, with the expectation of equal numbers of dominant and recessive characters in the offspring, and with a binomial distribution of sex ratios in a population of progenies. Among pairs of dizygotic human twins, one would expect about 25% to be both boys, 25% both girls, and 50% one of each.

This may well be the view held by a majority of the world's biologists who are not specialists in evolutionary theory. Most adherents of this theory know that the number of boys born does not exactly equal the number of girls, but they also know that the difference is slight and readily explained by sex-biased viability, slightly different X and Y chromosome mechanics, or other minor influences on an otherwise ideal Mendelian ratio.

The simple Mendelian model furnishes a useful set of contrasts for the evaluation of other theories. The expected binomial distribution of sex ratios among progenies of a certain size is especially useful as a null hypothesis.

A THEORY OF OPTIMIZATION IN RELATION TO PARENTAL RESOURCES

I accept the reasoning of Smith & Fretwell (1974) that optimum investment from an offspring viewpoint will be greater than that from a parental viewpoint but that, when investment (e.g. provisioning with yolk) is finished before the offspring has appreciable resources of its own the parent can be expected to win the conflict. Only with such late issues as weaning conflict can the young be expected to achieve any compromise (Trivers 1974). I also accept the reasoning of Trivers & Willard (1973) that sex differences in fitness variance may raise the parent's optimum expenditure for one sex (usually males) over the other.

It follows that offspring sex ratio may constrain parental investment. Suppose, for example, that the ideal investment is 1.1 units of resource in a son and 1.0 in a daughter, for a mammalian mother in a species with one or a few young in a litter. She can achieve an optimum investment in each offspring only if her total investment is 1.0, 1.1, 2.0, 2.1, etc. This constraint may conflict with optimization of the mother's total investment in relation to her own reproductive value, physiological capabilities, territorial resources, etc. A female that is better off than another in these respects should be inclined to devote a greater allocation of resources to reproduction. This optimum total investment should be broadly and continuously distributed among the mothers in a population. I would expect selection to favour a mother that attempts to rear that combination of sons and daughters that has a total optimum expenditure that approximates her own personal optimum.

I regard figure 1 as a realistic interpretation of offspring costs and maternal capabilities in man and other single-birth species. One might expect an autocorrelation in maternal resources from one conception to the next. A woman who is healthy and well fed at the conception of her nth child will most probably be similarly favoured at the $(n+1)$th. She should be a son specialist. A woman who is consistently below mean vigour should be a daughter specialist. The argument is given in more detail by Trivers and Willard (1973).

An average woman's capacity for motherhood might be strongly age dependent. As a teenager she might produce mostly daughters, then switch to sons during the years of peak capabilities in her twenties, then back to daughters later on. She might end up with the same number of boys as girls, but show an autocorrelation in the sexes of consecutive births, with a run of sons in mid-career. Both the variation among women and the variation over time would result in this sort of autocorrelation.

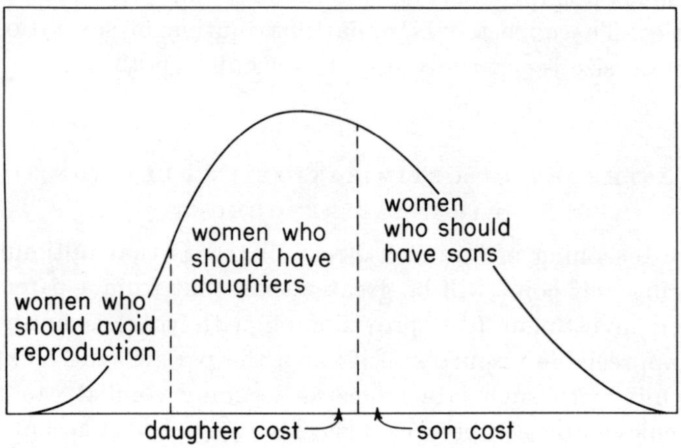

FIGURE 1. Adaptive choice of offspring sex in relation to maternal capability in man and other single-birth species.

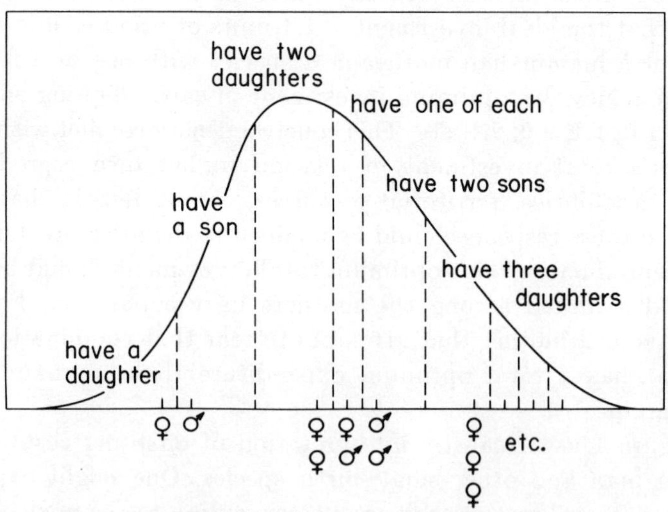

FIGURE 2. Litter sex-ratio optimization in a species that most often has two, but occasionally one or three, young in a litter or clutch.

Now consider a species that usually produces more than one young at a time, but few enough for litter sex ratio to be important for litter cost, again with an optimum of 1.1 cost units for a son and 1.0 for a daughter (figure 2). An approximate matching of maternal capability with litter cost would require not only regulation of litter size, but also of litter sex ratio. If the graphic model is realistic, certain observations would be expected with such forms as bears, hyaenas, and caciques in which the usual litter or clutch size is two. Modest cost differences between sons and daughters would imply a great predominance of single-sex litters; rare single births should be mostly sons; rare triplets mostly females. Litter size and sex ratio should be predictable from assessments of maternal age, physiology, and proprietary resources.

Infinite-population models

MacArthur (1965) extended Fisher's reasoning to the individual level by showing that an infinite population of parents that matched the evolutionary equilibrium with their own progenies would not be vulnerable to replacement by any other type of superior fitness. Spieth (1974, pp. 843, 844) has argued that MacArthur's reasoning does not really relate to the question of parental adaptation. It is also an incomplete argument, because it ignores Kolman's (1960) demonstration that in such deterministic models selection has no effect on the variance of sex ratio among progenies. Mutation pressure and genetic drift would make MacArthur's population vulnerable to an endless array of sex-ratio types. Selection would merely act to adjust mean sex ratio by favouring those types that overproduced the under-represented sex whenever the population deviated from its equilibrium. By Kolman's reasoning one would expect a human population to contain many more mostly male or mostly female progenies than would be expected from the Mendelian model, but this variation would be non-adaptive. Progeny sex biases would not be predictable from assessment of maternal capability.

Verner's finite-population model

Neither MacArthur's nor Kolman's reasoning is applicable to finite populations, which will always be subject to finite fluctuations in sex ratio. When one sex is temporarily in excess, producers of that sex are selected against and producers of the opposite sex favoured. Verner (1965) was the first to note an asymmetry in these positive and negative effects. The tendency to produce mostly one sex will lose more when that sex is over-abundant than it will gain when the opposite sex is in excess. This is because, in a finite population, a progeny sex ratio will have a finite effect on population sex ratio. An all-male progeny would itself reduce the advantage of a temporary female excess and aggravate the disadvantage of a male excess. Verner showed with some numerical examples that the optimum parental strategy, in a population in which sex ratio varies symmetrically but unpredictably about the evolutionary equilibrium, is to spend resources equally on sons and daughters. Pianka (1974) repeats Verner's idea with some more numerical examples. A more general and refined statement would be Charnov's (1975) equation (9) with the r replaced by a random variable with a mean of 0.5.

The more closely and reliably a parent can match the population equilibrium with its own progeny, the greater will be its average success in producing grandchildren, in any finite population. This theory gains credibility from the realization that it is not the sex ratio of the whole population that is the important environmental variable, but only that of the local group of potential mates. In sparse or viscous populations an individual may find only a few other members of its species in its search for a mate. For such an individual the environmental sex ratio can be most uncertain, and could deviate widely from the population mean. Selection for some kind of *sex-ratio homeostasis* would be most intense in this kind of population.

Verner's theory loses credibility for many species on realization that selection deals with the sex ratio of the young at the time of mating, and that this may come long after the end of any possible parental control. As an extreme example, consider a prolific marine fish that produces planktonic eggs that receive no attention after fertilization. What good would it do to have a mechanism that assured that a million sons were exactly balanced by a million daughters when survivors can be no more than a few individuals selected independently of sex? The Verner effect should have evolutionary consequences only in those species in which progeny sex ratio at the end of parental control has some influence on sex ratio at maturity.

Recent extensions of Verner's theory

Taylor & Sauer (1979) and Maynard Smith & Haigh (personal communication) analysed the Verner effect mathematically to determine how strong an evolutionary factor it might be. Their calculations concurred that sex-ratio homeostasis (alternating sons and daughters) would have an advantage of approximately $1/2N$ over Mendelian sex determination, where N is the number of individuals in the neighbourhood within which mates must be found. Taylor & Sauer also showed, given a genetically controlled dimorphism of Mendelian and homeostatic forms, that the homeostatic would have a density dependent fixation probability greater than that of the Mendelian.

There are several reasons why $1/2N$ may be a generous estimate of the force of selection: (1) it is accurate only when homeostasis is rare, and decreases as this condition becomes common; (2) it puts Mendelian sex determination in competition with fully developed homeostasis, and this implies greater fitness variance than would be expected of more plausible polygenic influences on sex ratio; (3) it assumes zero mortality between the end of investment or control by parents and the quest for mates by their young; (4) sex-ratio homeostasis is assumed to be cost free; and (5) there are no other ways of avoiding an unfavourable sex ratio, such as increased dispersal. These qualifications mostly mean that well developed homeostatic control of sex ratio might be expected only in species with young that remain under parental control almost to sexual maturity, and then disperse so little that only a few other conspecifics are likely to be encountered as potential mates. A search for sex-ratio homeostasis should concentrate on species that seem to have this kind of life history.

Taylor & Sauer (1980) used computer simulations to confirm the replacement of Mendelian sex determination by sex-ratio homeostasis. These simulations were free of some of the unrealistic simplifications that had to be made for mathematical analysis. For instance, both clutch size and population size varied in a plausible fashion. Homeostasis was made to depend on an autosomal gene inherited in fully stochastic Mendelian fashion. In their model populations, regulated at a mean of 20 individuals but subject to wide fluctuation, the gene for homeostasis was fixed in 56 % of 2000 runs, the gene for Mendelian sex determination in 44 %. Essentially the same results were produced when the homeostasis gene was dominant as when it was recessive. This agrees well with values (slightly different for dominants and recessives) predicted mathematically.

In one important respect, this study underestimates the effect of homeostasis. With only 20 individuals on the average, and with frequent bottlenecks much smaller than 20, genetic drift is a powerful force. A more realistic simulation, for many species, would be a simulataneous running of many small neighbourhoods (e.g. 20 individuals) with occasional migration from one to another. The small-neighbourhood advantage of homeostasis would be in full force, but the large total population would greatly reduce the competing force of genetic drift.

Canons of evidence

Variations in mean sex ratios among populations or their subgroups may sometimes be attributed to natural selection (see, for example, Howe 1977), but they are irrelevant to the question of parental adaptation. For any feature to be adaptive in my usage it must be a kind of local optimum, such that any slight change would result in reduced fitness. If a group of parents invests 60 % of its resources in sons and 40 % in daughters, the question of the adaptiveness of this mean investment ratio would depend entirely on the population equilibrium, and there is no way to determine this value. This uselessness of mean values is illustrated in table 1. If the observed mean is exactly what would be produced ultimately by natural selection acting on a specified subgroup of a population, we could say that this value is of equal fitness to any other, but it could never be a unique optimum and would sometimes be inferior to a real optimum at zero or unity. Evidence of adaptive sex ratio can come only from observation of sex-ratio variance among progenies. The theories reviewed above make specific predictions on how these variances should behave as a function of reproductive physiology or of demographic factors (table 2).

Man may be the best possible subject of investigation. The slightly greater male size at all stages of development, greater male variance in fitness in all extant or likely societies, and small litter size all provide conformity to requirements of the Trivers–Willard theory. The prolonged period of parental investment provides a long list of conceivable control mechanisms for regulating sex ratio as expected from Verner's theory: sexually selective fertilization, implantation, abortion or infanticide; sexually biased allocation of resources during childhood. Likewise any sex-ratio homeostasis would be fully effective because the time between the end

TABLE 1. IRRELEVANCE OF AN OBSERVED MEAN SEX RATIO TO CONCLUSIONS ON PARENTAL ADAPTATION

observed $p_{♂♂}$	equilibrium $p_{♂♂}$	optimum $p_{♂♂}$
0.6	0.5	0
0.6	0.6	none
0.6	0.7	1

TABLE 2. CONTRASTING EXPECTATIONS OF PROGENY SEX-RATIO THEORIES

theory	*expectations*
Mendelian	Every sex determination is an event independent of every other. For numerically equal sets of offspring, the population should show a binomial frequency distribution of sex ratio.
Trivers–Willard	Progeny sex ratio should be predictable from maternal condition in species with small litters and a son–daughter difference in cost. In the classic example, mothers in good condition should show a son bias, those in poor condition a daughter bias. This effect would be manifest in a progeny sex-ratio variance greater than the binomial, and should be missing from species that lack the required reproductive physiology.
Kolman	Sex-ratio variance among progenies greater than the binomial, but with no association between maternal resources and offspring sex ratio.
Verner	Sex-ratio homeostasis. Some mechanisms should act to provide a closer match between progeny sex ratio and population equilibrium than would be expected of chance sex determination. Variance among progenies should be less than the binomial. This effect should be present only in species that have small or viscous populations and minimal mortality between independence from parents and mating.

of investment by parents and the mating of offspring would be brief or non-existent in human societies. The mass and reliability of data for human populations are far greater than for any other.

Hunting-gathering and primitive agricultural populations would also conform to Verner's theory's requirement for limited choice of mate. Even after the originally sparse populations become denser, and after advances in transportation technology, social class structures would have continued to limit the size of the group in which a mate must be found. If Battisto had had a son and a daughter, instead of Blanche and Kate, he might have been more productive of grandchildren.

There remains the question of just how extreme an effect must be demanded before it can be recognized as evidence for one of the theories. We do not expect perfection in biological mechanisms, but we have come to expect precision. My reaction to an effect so minor that it takes a large mass of data merely to demonstrate its existence would be that the effect could scarcely represent a biological adaptation. If I were told that a centipede has exactly 50 right legs and were asked to predict how many left legs it would be found to have, I should not be inclined to use properties of the binomial distribution in my answer. Perhaps it is too much to ask of a centipede that a male–female homeostasis for its fertilized eggs be as precise as its left–right homeostasis for legs, but I should require a sex-ratio variance much less than the binomial before I would recognize evidence for Verner's theory.

Evidence from human populations

The evidence clearly supports the Mendelian theory and no other. Mean sex ratios of groups defined by race, age, parity, etc. may be slightly different, and this suggests that mean values may be subject to modification by selection, but there is no evidence of effects of selection on the variance. A study of 116 458 Mormon families failed to reject the independent-event hypothesis (Greenberg & White 1967). A similar study of 51 868 Finnish families showed an excess variance attributable to a correlation in sex of successive offspring, but the correlation in sex of successive offspring, but the correlation was almost certainly less than 0.02 (Edwards 1970). This is an extremely good approximation to a zero correlation. These are merely two of the most recent and comprehensive and methodologically enlightened studies, but there are many others. None gives any evidence of noteworthy effects on sex-ratio variance.

Man meets the requirements of both the Trivers–Willard and the Verner theories, which predict opposite effects on sex-ratio variance. It is possible that both effects are strong and exactly cancel. It seems exceedingly unlikely that they would exactly cancel in all populations studied, although it is conceivable that a relation might be demonstrable in primitive societies between sex-ratio variance and demographic structure as it affects mating choices.

Evidence from other mammals

James (1975, 1976) followed an intuitive impression of sex-ratio homeostasis in mammalian litters and reviewed information on many, mostly domesticated, species. Only the wild species are relevant here, a hare (63 litters), a rabbit (197 litters), a deer mouse (1139 litters), and newly domesticated mink (1539 litters). None deviated significantly from the binomial, and all may thereby be said to support the simple Mendelian model, although the hare data were too meagre to indicate any but extreme effects.

James compiled his information from all available sources for each species, and pointed out that variation in male–female probabilities between stocks could produce totals in which homeostasis within each stock would be masked. There was evidence that this was true for some of the domestic varieties. It remains clear that deviations from the binomial in at least three species from three different orders are too small to count as evidence for either theory of adaptive sex ratio.

Collard lemmings (Hasler & Banks 1975) and wood lemmings (Fredga *et al.* 1977; Bengtsson 1977) have strong and consistent female majorities (*ca.* 75%) in the populations studied. Sex-ratio variance among progenies is extreme in both species, with some females seemingly capable of producing only daughters but others (at least in the collard lemming) with significant overproduction of males. There is evidence in the wood lemming that the all-daughter females are of XY karyotype, with an X that is special in two ways: it causes its bearer to be female; and it produces mainly or exclusively X-bearing eggs that all give rise to daughters. Bengtsson showed that the equilibrium proportion of males in a population with this special X could be as low as 25%. His analysis is undoubtedly correct, but does not solve the problem of the female majority. It did not consider what I would regard as the important question of selection on autosomal modifiers that would cause the XY females to be males. In joining the minority sex they would have a large mating advantage that would rapidly increase the frequencies of the autosomal modifiers. Only by postulating that there can be no variation in autosomal sex determiners would Bengtsson's ratio be stable under outcrossing.

This deficiency in Bengtsson's analysis was pointed out by Stenseth (1978), who proposed that inbreeding may account for the wood lemming's great female majority. He reasoned that during times of sparse populations there may be mostly brother–sister and parent–offspring matings. Families that are mostly female could increase more rapidly than those with more males, and contribute descendants disproportionately to later dense populations. Bengtsson predicted that female biases will be found to characterize other species with widely fluctuating populations. The general relation between inbreeding and sex ratio was formulated by Hamilton (1967).

Of more immediate relevance to my topic is the high sex-ratio variance, apparently attributable to there being two kinds of X chromosome. The possibility should be considered that a cycle of inbreeding and outbreeding will select for a simple chromosomal polymorphism rather than polygenic mechanisms of female bias. There is no suggestion in any of the works cited that litter sex ratio is in any way related to maternal resources as would be expected from the Trivers–Willard theory. In the absence of any other information these rodents would seem to support Kolman's theory.

Other data are available but they are mostly even less adequate than the hare data cited by James. Herskovitz (1977), for example, listed litter sex ratios for marmosets. The usual litter size is two, so that the species is especially interesting in relation to the Trivers–Willard theory (figure 2). In 91 sets of twins, 24 were both male and 24 both female. This 52.3% same-sex pairs is close to the 50% expected of the binomial distribution. Unfortunately the 95% confidence interval for 0.523 reaches to about 0.63, and if this were the population mean, it might well be considered evidence for the Trivers–Willard theory or for Kolman's theory. Much larger samples will be necessary for discriminating among the theories being considered.

Other vertebrates

The best set of data for any bird, perhaps the best for any wild vertebrate generally, is that of Cooke and others (in preparation) on the snow goose colony at Churchill, Manitoba. They have clutch sex ratios on more than 1000 nests over a period of several years. The data support a strictly Mendelian interpretation and give no support to any theory of adaptive sex ratio. I know of no other published sex ratios on more than 65 clutches of a bird population, and such sample sizes are not worth considering. Some colleagues and I hope soon to publish data on some viviparous sharks and an embiotocid fish.

Conclusions

I find no support for any theory of adaptive sex ratio. Except in the lemmings, sex ratios of wild vertebrates show the binomial distributions to be expected if each sex determination is an independent event. Sex seems to be just another Mendelian unit character.

This conclusion does not, of course, mean that population sex ratio, or the sex ratio of defined groups within a population, is not subject to modification by natural selection. The seasonal variation in pooled sex ratios of grackles observed by Howe (1977) is plausibly based on varying population equilibria and attributable to evolved mechanisms for altering male–female probabilities at conception. The same may be true for the seasonal succession in mean litter sex ratios of seals noted by Trivers & Willard (1973) and the slight differences between races, social classes and age categories found in man. I am merely rejecting the possibility of saying, for any pair of parents, that if their allocation of investment to sons and daughters had been different from that observed, they would have been less biologically fit.

The evidence should raise doubts on theories of adaptive sex ratio in all outbred dioecious organisms, but undoubtedly there are legitimate instances of adaptive control of offspring sex ratio. Some of the examples discussed by Hamilton (1967) seem undeniable, for instance the production of many daughters plus one or two sons in species that reproduce solely by brother–sister mating. Perhaps all real examples of adaptive sex ratio will be seen to depend on close inbreeding and special cytological machinery.

Caution is also advisable on adaptive sex-ratio concepts for haplodiploid organisms. In particular, it is important to distinguish population equilibria from individual optima, concepts that will seldom coincide. For instance, it may be true, as Trivers & Hare (1976) maintain, that selection on queens in eusocial

Hymenoptera would, if unopposed, result in equal sex progenies, whereas selection on workers would produce progenies with three females to every male. These are hypothetical equilibria for populations, not optima for anyone. If, as would normally be ture, both kinds of selection have been acting, a compromise equilibrium would result. This means that, from a queen's viewpoint, there is a male excess, and from the workers' a female excess. So all-son progenies would always be optimal for queens, and all-daughter progenies for workers.

Remaining questions

I find it rather mysterious that adaptive control of progeny sex ratio seems not to have evolved. In particular the conformity of human progeny sex ratios to binomial distributions seems to contradict evolutionary theory. Either the physiological advantage of adjusting offspring sex to maternal capabilities, or the demographic advantage of decreasing competition for mates, ought to have produced noteworthy effects. Instead, deviations from random sex determination are trivial at best.

Perhaps this mystery is merely part of a greater one, the stability of Mendelian inheritance in general. Sex is only one of many offspring characters that would seem adaptive for a parent to control. For instance, in human populations affected by sickle-cell anaemia, it would be advantageous for a heterozygous woman to have her *A* eggs fertilized only by *a*-bearing sperm, and vice versa, or even to abort all homozygous embryos. Yet if mated to another heterozygote she will reliably submit to the Mendelian lottery, even though this means markedly lowered fitness for half her children. That parents have no adaptive control of autosomal recombination makes it less surprising that they have no control of sex-chromosome recombination.

Liberman (1976) has examined this general problem and found, discouragingly, reasons for believing that the machinery of Mendelian inheritance ought not to be stable in evolution. I have suggested (Williams 1979) that the machinery of mitosis and meiosis, with its exactly equal replication of each gene, and exactly equal sharing of resources, has the look of a stable conflict resolution. The really fundamental questions in evolution may be answerable only by regarding each gene as ultimately in conflict with every other gene, even those at other loci in the same cell. A really valid theory of natural selection must be based ultimately on selfish replicators, genes and all other entities capable of the biased accumulation of different variant forms, as Dawkins (1978) has urged.

A step in this direction with sex ratio might be to consider it explicitly in relation to parent–offspring conflict. Suppose there were a population that met the requirements of Verner's theory and that therefore had well developed sex-ratio homeostasis. We can imagine this to be achieved in a low-cost cytological manner, with the XX female producing two eggs at a time. Then if one is fertilized by an X-bearing sperm, she will allow the other to be fertilized only by a Y-bearing sperm, and vice versa.

Now introduce into the population a mutant gene that, in a sperm, at least occasionally results in its being able to overcome the mechanism of fertilization preference by an egg. This would mean the replacement of death by life for that gene. It could only mean a minor quantitative loss of fitness by the mother from her having two young of the same sex. Perhaps it is understandable that genes that would frustrate mechanisms of sex-ratio homeostasis would be favourably selected, even if the result were a decreased efficiency of reproduction.

References (Williams)

Bengtsson, B. O. 1977 Evolution of the sex ratio in the wood lemming, *Myopus schisticolor*. In *Measuring selection in natural populations* (ed. F. Christiansen & T. Fenchel), pp. 333–343. New York: Springer.

Charnov, E. L. 1975 Sex ratio selection in an age-structured population. *Evolution* **29**, 336–368.

Dawkins, R. 1978 Replicator selection and the extended phenotype. *Z. Tierpsychol.* **47**, 61–76.

Edwards, A. W. F. 1970 The search for genetic variability of the sex ratio. In *The biosocial aspects of sex* (ed. G. A. Harrison & J. Peel), pp. 55–60. Oxford: Blackwell.

Eshel, I. 1975 Selection on sex-ratio and the evolution of sex-determination. *Heredity, Lond.* **34**, 351–361.

Fisher, R. A. 1930 *The genetical theory of natural selection*. Oxford University Press.

Fredga, K., Gropp, A., Winking, H. & Frank, F. 1977 A hypothesis explaining the exception sex ratio in the wood lemming (*Myopus schisticolor*). *Hereditas* **85**, 101–104.

Greenberg, R. A. & White, C. 1967 The sexes of consecutive sibs in human sibships. *Hum. Biol.* **39**, 374–404.

Hamilton, W. D. 1967 Extraordinary sex ratios. *Science, N.Y.* **156**, 477–488.

Hasler, J. F. & Banks, E. M. 1975 Reproductive performance and growth in captive collard lemmings (*Dicrostonyx groenlaudicus*). *Can. J. Zool.* **53**, 777–787.

Hershkovitz, P. 1977 *Living new world monkeys* (*Platyrrhini*) *with an introduction to primates*, vol. 1. Chicago University Press.

Howe, H. F. 1977 Sex-ratio adjustment in the common grackle. *Science, N.Y.* **198**, 744–746.

James, W. H. 1975 The distributions of the combinations of the sexes in mammalian litters. *Genet. Res.* **26**, 45–53.

James, W. H. 1976 The combinations of the sexes in twin lambings. *Genet. Res.* **28**, 277–280.

Kolman, W. A. 1960 The mechanism of natural selection for the sex ratio. *Am. Nat.* **94**, 373–377.

Liberman, U. 1976 Modifier theory of meiotic drive: Is Mendelian segregation stable? *Theor. Popul. Biol.* **10**, 127–132.

MacArthur, R. A. 1965 Ecological consequences of natural selection. In *Theoretical and mathematical biology* (ed. T. H. Waterman & H. J. Morowitz), pp. 388–397. New York: Blaisdell.

Pianka, E. R. 1974 *Evolutionary ecology*. New York: Harper & Row.

Shaw, R. F. 1958 The theoretical genetics of the sex ratio. *Genetics, Princeton* **43**, 149–163.

Smith, C. C. & Fretwell, S. D. 1974 The optimum balance between size and number of offspring. *Am. Nat.* **108**, 499–506.

Spieth, P. T. 1974 Theoretical considerations of unequal sex ratios. *Am. Nat.* **108**, 837–849.

Stenseth, N. C. 1978 Is the female biased sex ratio in wood lemming *Myopus schisticolor* maintained by cyclic inbreeding? *Oikos* **30**, 83–89.

Taylor, P. D. & Sauer, A. 1980 The selective advantage of sex-ratio homeostasis. *Am. Nat.* (In the press.)

Trivers, R. L. 1974 Parent–offspring conflict. *Am. Zool.* **14**, 249–264.

Trivers, R. L. & Hare, H. 1976 Haplodiploidy and the evolution of the social insects. *Science, N.Y.* **191**, 249–263.

Trivers, R. L. & Willard, D. E. 1973 Natural selection of parental ability to vary the sex ratio of offspring. *Science, N.Y.* **179**, 90–92.

Verner, S. 1965 Selection for sex ratio. *Am. Nat.* **99**, 419–421.

Williams, G. C. 1979 Kin selection and the paradox of sexuality. In *American Association for the Advancement of Science Symposium: Sociobiology: Beyond nature–nurture* (ed. G. W. Barlow & J. Silverberg). (In the press.)

Wilson, E. O. 1975 *Sociobiology, the new synthesis*. Cambridge, Mass.: Belknap Press.

Proc. R. Soc. Lond. B **205**, 581–598 (1979)
Printed in Great Britain

The spandrels of San Marco and the Panglossian paradigm: a critique of the adaptationist programme

BY S. J. GOULD AND R. C. LEWONTIN
Museum of Comparative Zoology, Harvard University, Cambridge, Massachusetts 02138, *U.S.A.*

An adaptationist programme has dominated evolutionary thought in England and the United States during the past 40 years. It is based on faith in the power of natural selection as an optimizing agent. It proceeds by breaking an organism into unitary 'traits' and proposing an adaptive story for each considered separately. Trade-offs among competing selective demands exert the only brake upon perfection; non-optimality is thereby rendered as a result of adaptation as well. We criticize this approach and attempt to reassert a competing notion (long popular in continental Europe) that organisms must be analysed as integrated wholes, with *Baupläne* so constrained by phyletic heritage, pathways of development and general architecture that the constraints themselves become more interesting and more important in delimiting pathways of change than the selective force that may mediate change when it occurs. We fault the adaptationist programme for its failure to distinguish current utility from reasons for origin (male tyrannosaurs may have used their diminutive front legs to titillate female partners, but this will not explain *why* they got so small); for its unwillingness to consider alternatives to adaptive stories; for its reliance upon plausibility alone as a criterion for accepting speculative tales; and for its failure to consider adequately such competing themes as random fixation of alleles, production of non-adaptive structures by developmental correlation with selected features (allometry, pleiotropy, material compensation, mechanically forced correlation), the separability of adaptation and selection, multiple adaptive peaks, and current utility as an epiphenomenon of non-adaptive structures. We support Darwin's own pluralistic approach to identifying the agents of evolutionary change.

1. INTRODUCTION

The great central dome of St Mark's Cathedral in Venice presents in its mosaic design a detailed iconography expressing the mainstays of Christian faith. Three circles of figures radiate out from a central image of Christ: angels, disciples, and virtues. Each circle is divided into quadrants, even though the dome itself is radially symmetrical in structure. Each quadrant meets one of the four spandrels in the arches below the dome. Spandrels – the tapering triangular spaces formed by the intersection of two rounded arches at right angles (figure 1) – are necessary architectural by-products of mounting a dome on rounded arches. Each spandrel contains a design admirably fitted into its tapering space. An evangelist sits in the upper part flanked by the heavenly cities. Below, a man representing one of the four Biblical rivers (Tigris, Euphrates, Indus and Nile) pours water from a pitcher into the narrowing space below his feet.

The design is so elaborate, harmonious and purposeful that we are tempted to view it as the starting point of any analysis, as the cause in some sense of the surrounding architecture. But this would invert the proper path of analysis. The

FIGURE 1. One of the four spandrels of St Mark's; seated evangelist above, personification of river below.

system begins with an architectural constraint: the necessary four spandrels and their tapering triangular form. They provide a space in which the mosaicists worked; they set the quadripartite symmetry of the dome above.

Such architectural constraints abound and we find them easy to understand because we do not impose our biological biases upon them. Every fan vaulted ceiling must have a series of open spaces along the mid-line of the vault, where the sides of the fans intersect between the pillars (figure 2). Since the spaces must exist, they are often used for ingenious ornamental effect. In King's College Chapel in Cambridge, for example, the spaces contain bosses alternately embellished with the Tudor rose and portcullis. In a sense, this design represents an 'adaptation', but the architectural constraint is clearly primary. The spaces arise as a necessary by-product of fan vaulting; their appropriate use is a secondary effect. Anyone who tried to argue that the structure exists because the alternation of rose and portcullis makes so much sense in a Tudor chapel would be inviting the same ridicule that Voltaire heaped on Dr Pangloss: 'Things cannot be other than they

FIGURE 2. The ceiling of King's College Chapel.

are...Everything is made for the best purpose. Our noses were made to carry spectacles, so we have spectacles. Legs were clearly intended for breeches, and we wear them.' Yet evolutionary biologists, in their tendency to focus exclusively on immediate adaptation to local conditions, do tend to ignore architectural constraints and perform just such an inversion of explanation.

As a closer example, recently featured in some important biological literature on adaptation, anthropologist Michael Harner has proposed (1977) that Aztec human sacrifice arose as a solution to chronic shortage of meat (limbs of victims were often consumed, but only by people of high status). E. O. Wilson (1978) has used this explanation as a primary illustration of an adaptive, genetic predisposition for carnivory in humans. Harner and Wilson ask us to view an elaborate social system and a complex set of explicit justifications involving myth, symbol, and tradition as mere epiphenomena generated by the Aztecs as an unconscious rationalization masking the 'real' reason for it all: need for protein. But Sahlins (1978) has argued that human sacrifice represented just one part of an elaborate cultural fabric that, in its entirety, not only represented the material expression of Aztec cosmology, but also performed such utilitarian functions as the maintenance of social ranks and systems of tribute among cities.

We strongly suspect that Aztec cannibalism was an 'adaptation' much like evangelists and rivers in spandrels, or ornamented bosses in ceiling spaces: a secondary epiphenomenon representing a fruitful use of available parts, not a cause of the entire system. To put it crudely: a system developed for other reasons generated an increasing number of fresh bodies; use might as well be made of them. Why invert the whole system in such a curious fashion and view an entire culture as the epiphenomenon of an unusual way to beef up the meat supply. Spandrels do not exist to house the evangelists. (Moreover, as Sahlins argues, it is not even clear that human sacrifice was an adaptation at all. Human cultural practices can be orthogenetic and drive towards extinction in ways that Darwinian processes, based on genetic selection, cannot. Since each new monarch had to outdo his predecessor in even more elaborate and copious sacrifice, the practice was beginning to stretch resources to the breaking point. It would not have been the first time that a human culture did itself in. And, finally, many experts doubt Harner's premise in the first place (Ortiz de Montellano 1978). They argue that other sources of protein were not in short supply, and that a practice awarding meat only to privileged people who had enough anyway, and who used bodies so inefficiently (only the limbs were consumed, and partially at that) represents a mighty poor way to run a butchery.)

We deliberately chose non-biological examples in a sequence running from remote to more familiar: architecture to anthropology. We did this because the primacy of architectural constraint and the epiphenomenal nature of adaptation are not obscured by our biological prejudices in these examples. But we trust that the message for biologists will not go unheeded: if these had been biological systems, would we not, by force of habit, have regarded the epiphenomenal adaptation as primary and tried to build the whole structural system from it?

2. The adaptationist programme

We wish to question a deeply engrained habit of thinking among students of evolution. We call it the adaptationist programme, or the Panglossian paradigm. It is rooted in a notion popularized by A. R. Wallace and A. Weismann (but not, as we shall see, by Darwin) towards the end of the nineteenth century: the near omnipotence of natural selection in forging organic design and fashioning the best among possible worlds. This programme regards natural selection as so powerful and the constraints upon it so few that direct production of adaptation through its operation becomes the primary cause of nearly all organic form, function, and behaviour. Constraints upon the pervasive power of natural selection are recognized of course (phyletic inertia primarily among them, although immediate architectural constraints, as discussed in the last section, are rarely acknowledged). But they are usually dimissed as unimportant or else, and more frustratingly, simply acknowledged and then not taken to heart and invoked.

Studies under the adaptationist programme generally proceed in two steps:

(1) An organism is atomized into 'traits' and these traits are explained as structures optimally designed by natural selection for their functions. For lack of space, we must omit an extended discussion of the vital issue: 'what is a trait?' Some evolutionists may regard this as a trivial, or merely a semantic problem. It is not. Organisms are integrated entities, not collections of discrete objects. Evolutionists have often been led astray by inappropriate atomization, as D'Arcy Thompson (1942) loved to point out. Our favourite example involves the human chin (Gould 1977, pp. 381–382; Lewontin 1978). If we regard the chin as a 'thing', rather than as a product of interaction between two growth fields (alveolar and mandibular), then we are led to an interpretation of its origin (recapitulatory) exactly opposite to the one now generally favoured (neotenic).

(2) After the failure of part-by-part optimization, interaction is acknowledged via the dictum that an organism cannot optimize each part without imposing expenses on others. The notion of 'trade-off' is introduced, and organisms are interpreted as best compromises among competing demands. Thus, interaction among parts is retained completely within the adaptationist programme. Any suboptimality of a part is explained as its contribution to the best possible design for the whole. The notion that suboptimality might represent anything other than the immediate work of natural selection is usually not entertained. As Dr Pangloss said in explaining to Candide why he suffered from venereal disease: 'It is indispensable in this best of worlds. For if Columbus, when visiting the West Indies, had not caught this disease, which poisons the source of generation, which frequently even hinders generation, and is clearly opposed to the great end of Nature, we should have neither chocolate nor cochineal.' The adaptationist programme is truly Panglossian. Our world may not be good in an abstract sense, but it is the very best we could have. Each trait plays its part and must be as it is.

At this point, some evolutionists will protest that we are caricaturing their view of adaptation. After all, do they not admit genetic drift, allometry, and a variety of reasons for non-adaptive evolution? They do, to be sure, but we make a different point. In natural history, all possible things happen sometimes; you generally do not support your favoured phenomenon by declaring rivals impossible in theory. Rather, you acknowledge the rival, but circumscribe its domain of action so narrowly that it cannot have any importance in the affairs of nature. Then, you often congratulate yourself for being such an undogmatic and ecumenical chap. We maintain that alternatives to selection for best overall design have generally been relegated to unimportance by this mode of argument. Have we not all heard the catechism about genetic drift: it can only be important in populations so small that they are likely to become extinct before playing any sustained evolutionary role (but see Lande 1976).

The admission of alternatives in principle does not imply their serious consideration in daily practice. We all say that not everything is adaptive; yet, faced with an organism, we tend to break it into parts and tell adaptive stories as if trade-offs among competing, well designed parts were the only constraint upon perfection for each trait. It is an old habit. As Romanes complained about A. R. Wallace in 1900: 'Mr. Wallace does not expressly maintain the abstract impossibility of laws and causes other than those of utility and natural selection... Nevertheless, as he nowhere recognizes any other law or cause..., he practically concludes that, on inductive or empirical grounds, there *is* no such other law or cause to be entertained.'

The adaptationist programme can be traced through common styles of argument. We illustrate just a few; we trust they will be recognized by all:

(1) If one adaptive argument fails, try another. Zig-zag commissures of clams and brachiopods, once widely regarded as devices for strengthening the shell, become sieves for restricting particles above a given size (Rudwick 1964). A suite of external structures (horns, antlers, tusks) once viewed as weapons against predators, become symbols of intraspecific competition among males (Davitashvili 1961). The eskimo face, once depicted as 'cold engineered' (Coon *et al.* 1950), becomes an adaptation to generate and withstand large masticatory forces (Shea 1977). We do not attack these newer interpretations; they may all be right. We do wonder, though, whether the failure of one adaptive explanation should always simply inspire a search for another of the same general form, rather than a consideration of alternatives to the proposition that each part is 'for' some specific purpose.

(2) If one adaptive argument fails, assume that another must exist; a weaker version of the first argument. Costa & Bisol (1978), for example, hoped to find a correlation between genetic polymorphism and stability of environment in the deep sea, but they failed. They conclude (1978, pp. 132, 133): 'The degree of genetic polymorphism found would seem to indicate absence of correlation with the particular environmental factors which characterize the sampled area. The results suggest that the adaptive strategies of organisms belonging to different phyla are different.'

(3) In the absence of a good adaptive argument in the first place, attribute failure to imperfect understanding of where an organism lives and what it does. This is again an old argument. Consider Wallace on why all details of colour and form in land snails must be adaptive, even if different animals seem to inhabit the same environment (1899, p. 148): 'The exact proportions of the various species of plants, the numbers of each kind of insect or of bird, the peculiarities of more or less exposure to sunshine or to wind at certain critical epochs, and other slight differences which to us are absolutely immaterial and unrecognizable, may be of

the highest significance to these humble creatures, and be quite sufficient to require some slight adjustments of size, form, or colour, which natural selection will bring about.'

(4) Emphasize immediate utility and exclude other attributes of form. Fully half the explanatory information accompanying the full-scale Fibreglass *Tyrannosaurus* at Boston's Museum of Science reads: 'Front legs a puzzle: how *Tyrannosaurus* used its tiny front legs is a scientific puzzle; they were too short even to reach the mouth. They may have been used to help the animal rise from a lying position.' (We purposely choose an example based on public impact of science to show how widely habits of the adaptationist programme extend. We are not using glass beasts as straw men; similar arguments and relative emphases, framed in different words, appear regularly in the professional literature.) We don't doubt that *Tyrannosaurus* used its diminutive front legs for something. If they had arisen *de novo*, we would encourage the search for some immediate adaptive reason. But they are, after all, the reduced product of conventionally functional homologues in ancestors (longer limbs of allosaurs, for example). As such, we do not need an explicitly adaptive explanation for the reduction itself. It is likely to be a developmental correlate of allometric fields for relative increase in head and hindlimb size. This non-adaptive hypothesis can be tested by conventional allometric methods (Gould (1974) in general; Lande (1978) on limb reduction) and seems to us both more interesting and fruitful than untestable speculations based on secondary utility in the best of possible worlds. One must not confuse the fact that a structure is used in some way (consider again the spandrels, ceiling spaces and Aztec bodies) with the primary evolutionary reason for its existence and conformation.

3. Telling stories

> 'All this is a manifestation of the rightness of things, since if there is a volcano at Lisbon it could not be anywhere else. For it is impossible for things not to be where they are, because everything is for the best' (Dr Pangloss on the great Lisbon earthquake of 1755 in which up to 50000 people lost their lives).

We would not object so strenuously to the adaptationist programme if its invocation, in any particular case, could lead in principle to its rejection for want of evidence. We might still view it as restrictive and object to its status as an argument of first choice. But if it could be dismissed after failing some explicit test, then alternatives would get their chance. Unfortunately, a common procedure among evolutionists does not allow such definable rejection for two reasons. First, the rejection of one adaptive story usually leads to its replacement by another, rather than to a suspicion that a different kind of explanation might be required. Since the range of adaptive stories is as wide as our minds are fertile, new stories can always be postulated. And if a story is not immediately available, one can always plead temporary ignorance and trust that it will be forthcoming, as did Costa & Bisol (1978), cited above. Secondly, the criteria for acceptance of a story are so loose that many pass without proper confirmation. Often, evolutionists use *consistency* with natural selection as the sole criterion and consider their work done when they concoct a plausible story. But plausible stories can always be told. The key to historical research lies in devising criteria to identify proper explanations among the substantial set of plausible pathways to any modern result.

We have, for example (Gould 1978) criticized Barash's (1976) work on aggression in mountain bluebirds for this reason. Barash mounted a stuffed male near the nests of two pairs of bluebirds while the male was out foraging. He did this at the same nests on three occasions at 10 day intervals: the first before eggs were laid, the last two afterwards. He then counted aggressive approaches of the returning male towards both the model and the female. At time one, aggression was high towards the model and lower towards females but substantial in both nests. Aggression towards the model declined steadily for times two and three and plummeted to near zero towards females. Barash reasoned that this made evolutionary sense since males would be more sensitive to intruders before eggs were laid than afterwards (when they can have some confidence that their genes are inside). Having devised this plausible story, he considered his work as completed (1976, pp. 1099, 1100):

> 'The results are consistent with the expectations of evolutionary theory. Thus aggression toward an intruding male (the model) would clearly be especially advantageous early in the breeding season, when territories and nests are normally defended...The initial aggressive response to the mated female is also adaptive in that, given a situation suggesting a high probability of adultery (i.e. the presence of the model near the female) and assuming that replacement females are available, obtaining a new mate would enhance the fitness of males... The decline in male–female aggressiveness during incubation and fledgling stages could be attributed to the impossibility of being cuckolded after the eggs have been laid...The results are consistent with an evolutionary interpretation.'

They are indeed consistent, but what about an obvious alternative, dismissed without test by Barash? Male returns at times two and three, approaches the model, tests it a bit, recognizes it as the same phoney he saw before, and doesn't bother his female. Why not at least perform the obvious test for this alternative to a conventional adaptive story: expose a male to the model for the *first* time after the eggs are laid.

Since we criticized Barash's work, Morton *et al.* (1978) repeated it, with some variations (including the introduction of a female model), in the closely related eastern bluebird *Sialia sialis*. 'We hoped to confirm', they wrote, that Barash's conclusions represent 'a widespread evolutionary reality, at least within the genus *Sialia*. Unfortunately, we were unable to do so.' They found no 'anticuckoldry' behaviour at all: males never approached their females aggressively after testing the model at any nesting stage. Instead, females often approached the male model and, in any case, attacked female models more than males attacked male models.

'This violent response resulted in the near destruction of the female model after presentations and its complete demise on the third, as a female flew off with the model's head early in the experiment to lose it for us in the brush' (1978, p. 969). Yet, instead of calling Barash's selected story into question, they merely devise one of their own to render both results in the adaptationist mode. Perhaps, they conjecture, replacement females are scarce in their species and abundant in Barash's. Since Barash's males can replace a potentially 'unfaithful' female, they can afford to be choosy and possessive. Eastern bluebird males are stuck with uncommon mates and had best be respectful. They conclude: 'If we did not support Barash's suggestion that male bluebirds show anticuckoldry adaptations, we suggest that both studies still had "results that are consistent with the expectations of evolutionary theory" (Barash 1976, p. 1099), as we presume any careful study would.' But what good is a theory that cannot fail in careful study (since by 'evolutionary theory', they clearly mean the action of natural selection applied to particular cases, rather than the fact of transmutation itself).

4. The master's voice re-examined

Since Darwin has attained sainthood (if not divinity) among evolutionary biologists, and since all sides invoke God's allegiance, Darwin has often been depicted as a radical selectionist at heart who invoked other mechanisms only in retreat, and only as a result of his age's own lamented ignorance about the mechanisms of heredity. This view is false. Although Darwin regarded selection as the most important of evolutionary mechanisms (as do we), no argument from opponents angered him more than the common attempt to caricature and trivialize his theory by stating that it relied exclusively upon natural selection. In the last edition of the *Origin*, he wrote (1872, p. 395):

> 'As my conclusions have lately been much misrepresented, and it has been stated that I attribute the modification of species exclusively to natural selection, I may be permitted to remark that in the first edition of this work, and subsequently, I placed in a most conspicuous position – namely at the close of the Introduction – the following words: "I am convinced that natural selection has been the main, but not the exclusive means of modification." This has been of no avail. Great is the power of steady misinterpretation.'

Romanes, whose once famous essay (1900) on Darwin's pluralism versus the panselectionism of Wallace and Weismann deserves a resurrection, noted of this passage (1900, p. 5): 'In the whole range of Darwin's writings there cannot be found a passage so strongly worded as this: it presents the only note of bitterness in all the thousands of pages which he has published.' Apparently, Romanes did not know the letter Darwin wrote to *Nature* in 1880, in which he castigated Sir Wyville Thomson for caricaturing his theory as panselectionist (1880, p. 32):

> 'I am sorry to find that Sir Wyville Thomson does not understand the principle of natural selection...If he had done so, he could not have written the following sentence in the Introduction to the Voyage of the Challenger: "The character of the abyssal fauna refuses to give the least support to the theory which refers the evolution of species to extreme variation guided only by natural selection." This is a standard of criticism not uncommonly reached by theologians and metaphysicians when they write on scientific subjects, but is something new as coming from a naturalist...Can Sir Wyville Thomson name any one who has said that the evolution of species depends only on natural selection? As far as concerns myself, I believe that no one has brought forward so many observations on the effects of the use and disuse of parts, as I have done in my "Variation of Animals and Plants under Domestication"; and these observations were made for this special object. I have likewise there adduced a considerable body of facts, showing the direct action of external conditions on organisms.'

We do not now regard all of Darwin's subsidiary mechanisms as significant or even valid, though many, including direct modification and correlation of growth, are very important. But we should cherish his consistent attitude of pluralism in attempting to explain Nature's complexity.

5. A partial typology of alternatives to the adaptationist programme

In Darwin's pluralistic spirit, we present an incomplete hierarchy of alternatives to immediate adaptation for the explanation of form, function, and behaviour.

(1) No adaptation and no selection at all. At present, population geneticists are sharply divided on the question of how much genetic polymorphism within populations and how much of the genetic differences between species is, in fact, the result of natural selection as opposed to purely random factors. Populations are finite in size and the isolated populations that form the first step in the speciation process are often founded by a very small number of individuals. As a result of this restriction in population size, frequencies of alleles change by *genetic drift*, a kind of random genetic sampling error. The stochastic process of change in gene frequency by random genetic drift, including the very strong sampling process that goes on when a new isolated population is formed from a few immigrants, has several important consequences. First, populations and species will become genetically differentiated, and even fixed for different alleles at a locus in the complete absence of any selective force at all.

Secondly, alleles can become fixed in a population *in spite of natural selection*. Even if an allele is favoured by natural selection, some proportion of population, depending upon the product of population size N and selection intensity s, will become homozygous for the less fit allele because of genetic drift. If Ns is large this random fixation for unfavourable alleles is a rare phenomenon, but if selection coefficients are on the order of the reciprocal of population size ($Ns = 1$) or smaller, fixation for deleterious alleles is common. If many genes are involved in influencing a metric character like shape, metabolism or behaviour, then the intensity of selection on each locus will be small and Ns per locus may be small. As a result, many of the loci may be fixed for non-optimal alleles.

Thirdly, new mutations have a small chance of being incorporated into a population, even when selectively favoured. Genetic drift causes the immediate loss of most new mutations after their introduction. With a selection intensity s, a new favourable mutation has a probability of only $2s$ of ever being incorporated. Thus, one cannot claim that, eventually, a new mutation of just the right sort for some adaptive argument will occur and spread. 'Eventually' becomes a very long time if only one in 1000 or one in 10000 of the 'right' mutations that do occur ever get incorporated in a population.

(2) No adaptation and no selection on the part at issue; form of the part is a correlated consequence of selection directed elsewhere. Under this important category, Darwin ranked his 'mysterious' laws of the 'correlation of growth'. Today, we speak of pleiotropy, allometry, 'material compensation' (Rensch 1959, pp. 179–187) and mechanically forced correlations in D'Arcy Thompson's sense (1942; Gould 1971). Here we come face to face with organisms as integrated wholes, fundamentally not decomposable into independent and separately optimized parts.

Although allometric patterns are as subject to selection as static morphology itself (Gould 1966), some regularities in relative growth are probably not under immediate adaptive control. For example, we do not doubt that the famous 0.66 interspecific allometry of brain size in all major vertebrate groups represents a selected 'design criterion,' though its significance remains elusive (Jerison 1973). It is too repeatable across too wide a taxonomic range to represent much else than a series of creatures similarly well designed for their different sizes. But another common allometry, the 0.2 to 0.4 intraspecific scaling among homeothermic adults differing in body size, or among races within a species, probably does not require a selectionist story though many, including one of us, have tried to provide one (Gould 1974). R. Lande (personal communication) has used the experiments of Falconer (1973) to show that selection upon *body size alone* yields a brain–body slope across generations of 0.35 in mice.

More compelling examples abound in the literature on selection for altering the timing of maturation (Gould 1977). At least three times in the evolution of arthropods (mites, flies and beetles), the same complex adaptation has evolved, apparently for rapid turnover of generations in strongly r-selected feeders on superabundant but ephemeral fungal resources: females reproduce as larvae and grow the next generation within their bodies. Offspring eat their mother from inside and emerge from her hollow shell, only to be devoured a few days later by their own progeny. It would be foolish to seek adaptive significance in paedomorphic morphology *per se*; it is primarily a by-product of selection for rapid cycling of generations. In more interesting cases, selection for small size (as in animals of the interstitial fauna) or rapid maturation (dwarf males of many crustaceans) has occurred by progenesis (Gould 1977, pp. 324–336), and descendant adults contain a mixture of ancestral juvenile and adult features. Many biologists have been tempted to find primary adaptive meaning for the mixture, but it probably arises as a by-product of truncated maturation, leaving some features 'behind' in the larval state, while allowing others, more strongly correlated with sexual maturation, to retain the adult configuration of ancestors.

(3) The decoupling of selection and adaptation.

(i) Selection without adaptation. Lewontin (1979) has presented the following hypothetical example: 'A mutation which doubles the fecundity of individuals will sweep through a population rapidly. If there has been no change in efficiency of resource utilization, the individuals will leave no more offspring than before, but simply lay twice as many eggs, the excess dying because of resource limitation. In what sense are the individuals or the population as a whole better adapted than before? Indeed, if a predator on immature stages is led to switch to the species now that immatures are more plentiful, the population size may actually decrease as a consequence, yet natural selection at all times will favour individuals with higher fecundity.'

(ii) Adaptation without selection. Many sedentary marine organisms, sponges and corals in particular, are well adapted to the flow régimes in which they live. A wide spectrum of 'good design' may be purely phenotypic in origin, largely induced by the current itself. (We may be sure of this in numerous cases, when genetically identical individuals of a colony assume different shapes in different microhabitats.) Larger patterns of geographic variation are often adaptive and purely phenotypic as well. Sweeney & Vannote (1978), for example, showed that many hemimetabolous aquatic insects reach smaller adult size with reduced fecundity when they grow at temperatures above and below their optima. Coherent, climatically correlated patterns in geographic distribution for these insects – so often taken as *a priori* signs of genetic adaptation – may simply reflect this phenotypic plasticity.

'Adaptation' – the good fit of organisms to their environment – can occur at three hierarchical levels with different causes. It is unfortunate that our language has focused on the common result and called all three phenomena 'adaptation': the differences in process have been obscured and evolutionists have often been misled to extend the Darwinian mode to the other two levels as well. First, we have what physiologists call 'adaptation': the phenotypic plasticity that permits organisms to mould their form to prevailing circumstances during ontogeny. Human 'adaptations' to high altitude fall into this category (while others, like resistance of sickling heterozygotes to malaria, are genetic and Darwinian). Physiological adaptations are not heritable, though the capacity to develop them presumably is. Secondly, we have a 'heritable' form of non-Darwinian adaptation in humans (and, in rudimentary ways, in a few other advanced social species):

cultural adaptation (with heritability imposed by learning). Much confused thinking in human sociobiology arises from a failure to distinguish this mode from Darwinian adaptation based on genetic variation. Finally, we have adaptation arising from the conventional Darwinian mechanism of selection upon genetic variation. The mere existence of a good fit between organism and environment is insufficient evidence for inferring the action of natural selection.

(4) Adaptation and selection but no selective basis for differences among adaptations. Species of related organisms, or subpopulations within a species, often develop different adaptations as solutions to the same problem. When 'multiple adaptive peaks' are occupied, we usually have no basis for asserting that one solution is better than another. The solution followed in any spot is a result of history; the first steps went in one direction, though others would have led to adequate prosperity as well. Every naturalist has his favourite illustration. In the West Indian land snail *Cerion*, for example, populations living on rocky and windy coasts almost always develop white, thick and relatively squat shells for conventional adaptive reasons. We can identify at least two different developmental pathways to whiteness from the mottling of early whorls in all *Cerion*, two paths to thickened shells and three styles of allometry leading to squat shells. All 12 combinations can be identified in Bahamian populations, but would it be fruitful to ask why – in the sense of optimal design rather than historical contingency – *Cerion* from eastern Long Island evolved one solution, and *Cerion* from Acklins Island another?

(5) Adaptation and selection, but the adaptation is a secondary utilization of parts present for reasons of architecture, development or history. We have already discussed this neglected subject in the first section on spandrels, spaces and cannibalism. If blushing turns out to be an adaptation affected by sexual selection in humans, it will not help us to understand why blood is red. The immediate utility of an organic structure often says nothing at all about the reason for its being.

6. Another, and unfairly maligned, approach to evolution

In continental Europe, evolutionists have never been much attracted to the Anglo-American penchant for atomizing organisms into parts and trying to explain each as a direct adaptation. Their general alternative exists in both a strong and a weak form. In the strong form, as advocated by such major theorists as Schindewolf (1950), Remane (1971), and Grassé (1977), natural selection under the adaptationist programme can explain superficial modifications of the *Bauplan* that fit structure to environment: why moles are blind, giraffes have long necks, and ducks webbed feet, for example. But the important steps of evolution, the construction of the *Bauplan* itself and the transition between *Baupläne*, must involve some other unknown, and perhaps 'internal', mechanism. We believe that English biologists have been right in rejecting this strong form as close to an appeal to mysticism.

But the argument has a weaker – and paradoxically powerful – form that has not been appreciated, but deserves to be. It also acknowledges conventional selection for superficial modifications of the *Bauplan*. It also denies that the adaptationist programme (atomization plus optimizing selection on parts) can do much to explain *Baupläne* and the transitions between them. But it does not therefore resort to a fundamentally unknown process. It holds instead that the basic body plans of organisms are so integrated and so replete with constraints upon adaptation (categories 2 and 5 of our typology) that conventional styles of selective arguments can explain little of interest about them. It does not deny that change, when it occurs, may be mediated by natural selection, but it holds that constraints restrict possible paths and modes of change so strongly that the constraints themselves become much the most interesting aspect of evolution.

Rupert Riedl, the Austrian zoologist who has tried to develop this thesis for English audiences (1977 and 1975, now being translated into English by R. Jefferies), writes:

> 'The living world happens to be crowded by universal patterns of organization which, most obviously, find no direct explanation through environmental conditions or adaptive radiation, but exist primarily through universal requirements which can only be expected under the systems conditions of complex organization itself... This is not self-evident, for the whole of the huge and profound thought collected in the field of morphology, from Goethe to Remane, has virtually been cut off from modern biology. It is not taught in most American universities. Even the teachers who could teach it have disappeared.'

Constraints upon evolutionary change may be ordered into at least two categories. All evolutionists are familiar with *phyletic* constraints, as embodied in Gregory's classic distinction (1936) between habitus and heritage. We acknowledge a kind of phyletic inertia in recognizing, for example, that humans are not optimally designed for upright posture because so much of our *Bauplan* evolved for quadrupedal life. We also invoke phyletic constraint in explaining why no molluscs fly in air and no insects are as large as elephants.

Developmental constraints, a subcategory of phyletic restrictions, may hold the most powerful rein of all over possible evolutionary pathways. In complex organisms, early stages of ontogeny are remarkably refractory to evolutionary change, presumably because the differentiation of organ systems and their integration into a functioning body is such a delicate process, so easily derailed by early errors with accumulating effects. Von Baer's fundamental embryological laws (1828) represent little more than a recognition that early stages are both highly conservative and strongly restrictive of later development. Haeckel's biogenetic law, the primary subject of late nineteenth century evolutionary biology, rested upon a misreading of the same data (Gould 1977). If development occurs in integrated packages, and cannot be pulled apart piece by piece in evolution, then the adaptationist programme cannot explain the alteration of developmental programmes underlying nearly all changes of *Bauplan*.

The German palaeontologist A. Seilacher, whose work deserves far more attention than it has received, has emphasized what he calls '*bautechnischer*', or *architectural*, constraints (Seilacher 1970). These arise not from former adaptations retained in a new ecological setting (phyletic constraints as usually understood), but as architectural restrictions that never were adaptations, but rather the necessary consequences of materials and designs selected to build basic *Baupläne*. We devoted

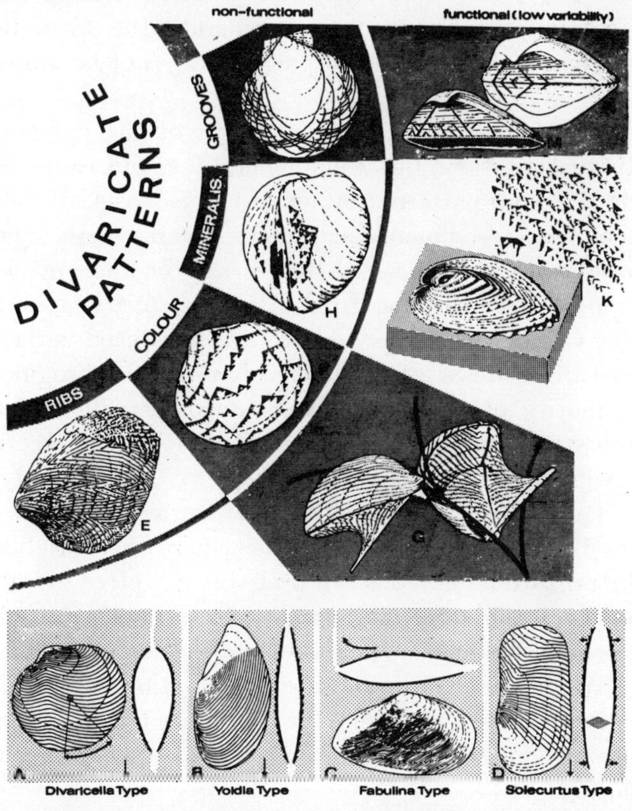

FIGURE 3. The range of divaricate patterns in molluscs. E, F, H, and L are non-functional in Seilacher's judgement. A–D are functional ribs (but these are far less common than non-functional ribs of the form E). G is the mimetic *Arca zebra*. K is *Corculum*. See text for details.

the first section of this paper to non-biological examples in this category. Spandrels must exist once a blueprint specifies that a dome shall rest on rounded arches. Architectural constraints can exert a far-ranging influence upon organisms as well. The subject is full of potential insight because it has rarely been acknowledged at all.

In a fascinating example, Seilacher (1972) has shown that the divaricate form of architecture (figure 3) occurs again and again in all groups of molluscs, and in brachiopods as well. This basic form expresses itself in a wide variety of structures: raised ornamental lines (not growth lines because they do not conform to the mantle margin at any time), patterns of coloration, internal structures in the mineralization of calcite, and incised grooves. He does not know what generates this pattern and feels that traditional and nearly exclusive focus on the adaptive value of each manifestation has diverted attention from questions of its genesis in growth and also prevented its recognition as a general phenomenon. It must arise from some characteristic pattern of inhomogeneity in the growing mantle, probably from the generation of interference patterns around regularly spaced centres; simple computer simulations can generate the form in this manner (Waddington & Cowe 1969). The general pattern may not be a direct adaptation at all.

Seilacher then argues that most manifestations of the pattern are probably non-adaptive. His reasons vary, but seem generally sound to us. Some are based on field observations: colour patterns that remain invisible because clams possessing them either live buried in sediments or remain covered with a periostracum so thick that the colours cannot be seen. Others rely on more general principles: presence only in odd and pathological individuals, rarity as a developmental anomaly, excessive variability compared with much reduced variability when the same general structure assumes a form judged functional on engineering grounds.

In a distinct minority of cases, the divaricate pattern becomes functional in each of the four categories (figure 3). Divaricate ribs may act as scoops and anchors in burrowing (Stanley 1970), but they are not properly arranged for such function in most clams. The colour chevrons are mimetic in one species (*Pteria zebra*) that lives on hydrozoan branches; here the variability is strongly reduced. The mineralization chevrons are probably adaptive in only one remarkable creature, the peculiar bivalve *Corculum cardissa* (in other species, they either appear in odd specimens or only as post-mortem products of shell erosion). This clam is uniquely flattened in an anterio-posterior direction. It lies on the substrate, posterior up. Distributed over its rear end are divaricate triangles of mineralization. They are translucent, while the rest of the shell is opaque. Under these windows dwell endosymbiotic algae!

All previous literature on divaricate structure has focused on its adaptive significance (and failed to find any in most cases). But Seilacher is probably right in representing this case as the spandrels, ceiling holes and sacrificed bodies of our first section. The divaricate pattern is a fundamental architectural constraint. Occasionally, since it is there, it is used to beneficial effect. But we cannot understand the pattern or its evolutionary meaning by viewing these infrequent and secondary adaptations as a reason for the pattern itself.

Galton (1909, p. 257) contrasted the adaptationist programme with a focus on constraints and modes of development by citing a telling anecdote about Herbert Spencer's fingerprints:

> 'Much has been written, but the last word has not been said, on the rationale of these curious papillary ridges; why in one man and in one finger they form whorls and in another loops. I may mention a characteristic anecdote of Herbert

Spencer in connection with this. He asked me to show him my Laboratory and to take his prints, which I did. Then I spoke of the failure to discover the origin of these patterns, and how the fingers of unborn children had been dissected to ascertain their earliest stages, and so forth. Spencer remarked that this was beginning in the wrong way; that I ought to consider the purpose the ridges had to fulfil, and to work backwards. Here, he said, it was obvious that the delicate mouths of the sudorific glands required the protection given to them by the ridges on either side of them, and therefrom he elaborated a consistent and ingenious hypothesis at great length. I replied that his arguments were beautiful and deserved to be true, but it happened that the mouths of the ducts did not run in the valleys between the crests, but along the crests of the ridges themselves.

We feel that the potential rewards of abandoning exclusive focus on the adaptationist programme are very great indeed. We do not offer a council of despair, as adaptationists have charged; for non-adaptive does not mean non-intelligible. We welcome the richness that a pluralistic approach, so akin to Darwin's spirit, can provide. Under the adaptationist programme, the great historic themes of developmental morphology and *Bauplan* were largely abandoned; for if selection can break any correlation and optimize parts separately, then an organism's integration counts for little. Too often, the adaptationist programme gave us an evolutionary biology of parts and genes, but not of organisms. It assumed that all transitions could occur step by step and underrated the importance of integrated developmental blocks and pervasive constraints of history and architecture. A pluralistic view could put organisms, with all their recalcitrant, yet intelligible, complexity, back into evolutionary theory.

References (Gould & Lewontin)

Baer, K. E. von 1828 *Entwicklungsgeschichte der Tiere.* Königsberg: Bornträger.

Barash, D. P. 1976 Male response to apparent female adultery in the mountain bluebird: an evolutionary interpretation. *Am. Nat.* **110**, 1097–1101.

Coon, C. S., Garn, S. M. & Birdsell, J. B. 1950 *Races.* Springfield, Ohio: C. Thomas.

Costa, R. & Bisol, P. M. 1978 Genetic variability in deep-sea organisms. *Biol. Bull.* **155**, 125–133.

Darwin, C. 1872 *The origin of species.* London: John Murray.

Darwin, C. 1880 Sir Wyville Thomson and natural selection. *Nature, Lond.* **23**, 32.

Davitashvili, L. S. 1961 *Teoriya polovogo otbora* [Theory of sexual selection]. Moscow: Akademii Nauk.

Falconer, D. S. 1973 Replicated selection for body weight in mice. *Genet. Res.* **22**, 291–321.

Galton, F. 1909 *Memories of my life.* London: Methuen.

Gould, S. J. 1966 Allometry and size in ontogeny and phylogeny. *Biol. Rev.* **41**, 587–640.

Gould, S. J. 1971 D'Arcy Thompson and the science of form. *New Literary Hist.* **2**(2), 229–258.

Gould, S. J. 1974 Allometry in primates, with emphasis on scaling and the evolution of the brain. In *Approaches to primate paleobiology. Contrib. Primatol.* **5**, 244–292.

Gould, S. J. 1977 *Ontogeny and phylogeny.* Cambridge, Mass.: Belknap Press.

Gould, S. J. 1978 Sociobiology: the art of storytelling. *New Scient.* **80**, 530–533.

Grassé, P.-P. 1977 *Evolution of living organisms*. New York: Academic Press.
Gregory, W. K. 1936 Habitus factors in the skeleton of fossil and recent mammals. *Proc. Am. phil. Soc.* **76**, 429–444.
Harner, M. 1977 The ecological basis for Aztec sacrifice. *Am. Ethnologist* **4**, 117–135.
Jerison, H. J. 1973 *Evolution of the brain and intelligence*. New York: Academic Press.
Lande, R. 1976 Natural selection and random genetic drift in phenotypic evolution. *Evolution* **30**, 314–334.
Lande, R. 1978 Evolutionary mechanisms of limb loss in tetrapods. *Evolution* **32**, 73–92.
Lewontin, R. C. 1978 Adaptation. *Scient. Am.* **239** (3), 156–169.
Lewontin, R. C. 1979 Sociobiology as an adaptationist program. *Behav. Sci.* (In the press.)
Morton, E. S., Geitgey, M. S. & McGrath, S. 1978 On bluebird 'responses to apparent female adultery'. *Am. Nat.* **112**, 968–971.
Ortiz de Montellano, B. R. 1978 Aztec cannibalism: an ecological necessity? *Science N.Y.* **200**, 611–617.
Remane, A. 1971 *Die Grundlagen des natürlichen Systems der vergleichenden Anatomie und der Phylogenetik*. Königstein-Taunus: Koeltz.
Rensch, B. 1959 *Evolution above the species level*. New York: Columbia University Press.
Riedl, R. 1975 *Die Ordnung des Lebendigen*. Hamburg: Paul Parey.
Riedl, R. 1977 A systems-analytical approach to macro-evolutionary phenomena. *Q. Rev. Biol.* **52**, 351–370.
Romanes, G. J. 1900 The Darwinism of Darwin and of the post-Darwinian schools. In *Darwin, and after Darwin*, vol. 2, new edn. London: Longmans, Green & Co.
Rudwick, M. J. S. 1964 The function of zig-zag deflections in the commissures of fossil brachiopods. *Palaeontology* **7**, 135–171.
Sahlins, M. 1978 Culture as protein and profit. *New York review of books*, 23 Nov., pp. 45–53.
Schindewolf, O. H. 1950 *Grundfragen der Paläontologie*. Stuttgart: Schweizerbart.
Seilacher, A. 1970 Arbeitskonzept zur Konstruktionsmorphologie. *Lethaia* **3**, 393–396.
Seilacher, A. 1972 Divaricate patterns in pelecypod shells. *Lethaia* **5**, 325–343.
Shea, B. T. 1977 Eskimo craniofacial morphology, cold stress and the maxillary sinus. *Am. J. phys. Anthrop.* **47**, 289–300.
Stanley, S. M. 1970 Relation of shell form to life habits in the Bivalvia (Mollusca). *Mem. geol. Soc. Am.* no. 125, 296 pp.
Sweeney, B. W. & Vannote, R. L. 1978 Size variation and the distribution of hemimetabolous aquatic insects: two thermal equilibrium hypotheses. *Science, N.Y.* **200**, 444–446.
Thompson, D. W. 1942 *Growth and form*. New York: Macmillan.
Waddington, C. H. & Cowe, J. R. 1969 Computer simulation of a molluscan pigmentation pattern. *J. theor. Biol.* **25**, 219–225.
Wallace, A. R. 1899 *Darwinism*. London: Macmillan.
Wilson, E. O. 1978 *On human nature*. Cambridge, Mass.: Harvard University Press.

Proc. R. Soc. Lond. B **205**, 599–604 (1979)
Printed in Great Britain

Introduction to general discussion

By A. J. Cain
Department of Zoology, University of Liverpool, Liverpool, U.K.

Having heard the papers given at this discussion meeting, one must agree that, whether or not one accepts all details of every statement made, there has been in the last 20 years a vast advance in evolutionary comprehension, associated particularly with advances in molecular biology, the theory of group selection, and the application of games theory to much of animal behaviour. On the one hand, our understanding of the nature of the inception of genetic variation has increased out of all recognition; on the other, numbers of phenomena which formerly were the subject of oleographed colour-plates showing 'bizarre habits' or 'outlandish structures' in popular natural histories are now brought into general theories which give us the means of asking answerable questions about them, and even making predictions. All of this greatly extends the scope of evolutionary theory.

There is, however, one major field of enquiry not mentioned by Maynard Smith – rightly, since it has had as yet very little influence on evolutionary theory – and that is natural selection as it actually goes on in the wild. Much of the work we have heard about at this meeting implies natural selection. If it can be shown that of different behavioural strategies directed to the same goal one is more efficient than the others, in expenditure of energy, use of a limited resource, or in resistance to the introduction of other strategies, then the others are less well adapted in these respects and may be expected to be selected out. In general, any demonstration of function implies the possibility of malfunction and the probability of selection acting to keep that function efficient. In some behavioural situations, the implied selection may be intense. A cock robin that is not sufficiently aggressive will not hold a territory and will not breed; but equally, if it is a little too aggressive and drives off both cocks and hens, it will hold a territory but will still not breed. Similarly, in Tinbergen's remarkable study of eggshell removal in gulls, taking too long away from the nest in removing the shell may leave the chick a prey to neighbour gulls. But staying to protect them and leaving the broken shell in the nest may mean that it is visible to predators, who then kill the chick. In such examples as these, selection against inappropriate behaviour may be from 50 to 100 %, and no population geneticist could doubt the efficacy of such a strength.

Nevertheless, the title of this meeting is 'The evolution of adaptation by means of natural selection' and one would be glad to see more positive knowledge about the actual degree of importance of natural selection in the wild. The researches of R. A. Fisher, J. B. S. Haldane, and Sewall Wright laid the foundations of the mathematics of population genetics. Even before their work, no one doubted that not everything we see in the living world was perfectly adapted to its purpose. Some variations may be due to pure chance (whatever that means), others may be the result of historical events, some adaptive phenomena may be continually altering as conditions change and never therefore perfect at any one time, some biochemical phenomena may be all-or-nothing because of their quantal nature and not gradually adaptable, and so on. What they did was to provide a mathematical theory covering all possible contingencies, from which quantitative predictions of both deterministic and stochastic processes could be made. That was a great gain, but it does not tell us what of all these possibilities are actually exemplified in the wild – what, in short, are relevant to the actual process of evolution.

It is not difficult to see why not. The available literature on actual genetic processes within wild populations is as yet very small, and increases only slowly. John Currey and I have just completed 16 years on one population of snails at Fyfield Down, and we seem to have seen only a single population cycle through. Presumably we have to do another 16 years to show that the very interesting genetic results are repeated. Other animals have a much more rapid turnover, but often they are ecologically intractable; *Drosophila*, for example, has four different ecologies in a single lifespan (for the egg, larva, pupa and adult), about any of which very little is known. It is not surprising that Dobzhansky used $\overline{\omega}$, the only measure of adaptedness that he could apply. In selection studies, we are many years from the situation in behavioural studies, in which one can think up some quirk of the selfish gene and confirm our predictions from a week's study of the excellent literature. There are many practical reasons (Cain 1977) why studies on actual selection on particular characters or character-complexes in real populations are difficult. I suspect that in the U.S.A. the system of funding research, which virtually demands that only work be started which promises results within 2 years, prevents anything but brief raids into the field and quick laboratory experiments from being carried out.

Moreover, most populations are not of very large numbers of individuals, and are spread over areas likely to be heterogeneous for many environmental characters. Mere sampling error will therefore usually prevent one from detecting selection coefficients of only 1–5%, yet in large populations such a strength is important enough to account for a great deal of genetic variation. We cannot, therefore, as a rule show that a given genetic phenomenon in the wild is unaffected by selection, although if large selection coefficients are acting we may be able to detect them. What we can do, however, is turn human frailty to account. Various excellent biologists at various times have pointed to various characters as trivial, or neutral, or unimportant to their possessors. In every case, when such characters have been properly investigated, quite large selection pressures have been found to be acting, or functions have been discovered which imply them (Cain 1964). Such judgements are in fact useless. I pointed out many years ago (Cain 1951 *a*, *b*, 1964) that all examples from the wild so far brought forward in which the variations 'must' be neutral or non-adaptive, or due to chance or history, or, indeed, are neutral in themselves but pleiotropes of, or linked to, something which is selected for, are properly to be described as examples about which not enough is known to come to a conclusion either way; the passage of time has brought nothing to alter that conclusion. To explain such examples on current theory is in fact merely to explain them away, and adds nothing to positive science. All that it does is to deter people from analysing them further, since if they are merely due to chance, then all is said that can be said about them, and there is nothing more one can do. Methodologically such an attitude is no more useful than vitalism or mysticism, and gets in the way of an objective assessment of the situation.

Lewontin (1974) has assured us that the 'textbook myth' of the objectivity of scientists 'has no congruence with reality. Long before there is any direct evidence, scientific workers have brought to the issue deep-seated prejudices; the more important the issue and the more ambiguous the evidence, the more important are the prejudices, and the greater the likelihood that two diametrically opposed and irreconcilable schools will appear...schools of thought about unresolved problems do not derive from idiosyncratic intuitions but from deep ideological biases reflecting social and intellectual world views.' No doubt, as we are wont to do, Lewontin is generalizing his own practice; nevertheless it is no news that no one achieves a completely objective view of anything. But the whole distinction of science (as against philosophy, for example) is that by the conflict of opinion *plus* an insistence on confirmation by experiment and directed observation at least by others, eventually problems do get solved in a more general sense than they were before. There is no other human activity of which this can be said; weak, slow and often dubious though it may be, scientific investigation is the only way by which we can reject proposed explanations and adjust ourselves to the world as it is. In his review of the *Essays presented to E. B. Ford*, Lewontin (1972) characterized the sort of determination of actual selection coefficients which is the only way of accumulating evidence on the importance of natural selection as a British upper middle class activity. Presumably when prejudice is strong, facts can be dispensed with as well: my own background and upbringing could only be distinguished by the extreme purist from working-class. Lewontin's statement is irrelevant; the only point scientifically is how far this activity stands up to scientific, not sociological, investigation. If not, then Lewontin's own conclusions require sociological investigation before we can accept them.

Ever since natural selection appeared on the scene, there have been those who voiced an *a priori* and dogmatic dislike of it. One classic document is George Bernard Shaw's preface to *Back to Methuselah*; for example, 'as compared with the open-eyed intelligent wanting and trying of Lamarck, the Darwinian process may be described as a chapter of accidents. As such, it seems simple, because you do not at first realize all that it involves. But when its whole significance dawns on you, your heart sinks into a heap of sand within you. There is a hideous fatalism about it, a ghastly and damnable reduction of beauty and intelligence, of strength and purpose, of honor and aspiration, to such casually picturesque changes as an avalanche may make in a mountain landscape, or a railway accident in a human figure.' This is an excellent statement of emotion; others, whether moved by that emotion or not, have objected to natural selection as in literal conflict with the words of some authoritative dogma or other, whether religious or political (if indeed one can make a distinction). I suspect from my own work that natural selection may have been very much more important than anyone has realized up to now. If so, can these emotional and other rejections of it, or, more generally, the tendency of the human race to take a non-objective view of evolution and kindred topics, be explained by natural selection? There is a possible evolutionary explanation, as yet untested, and no other scientific one that I know of.

One of the most revealing accounts of the life history in *Homo sapiens* was made quite unwittingly by Lorenz in his famous *King Solomon's ring*. There he describes the results of his successful breeding of water-shrews. The young, in the snug nest, looked after by their parents, grow rapidly, play vigorously, exercising themselves and each other, and eventually (whether by their own drives, or their parents', or more likely both) they break away from their childhood life and go wandering out into the world to find places for themselves. They are at this stage, adventurous, eager, inquisitive – typical adolescents. Of course, as the population remains stable, a vast percentage of them die since their efforts are unsuccessful. That is the natural fate of the majority of adolescents. Many of our social troubles at the present day stem from their survival in bulk in human populations and the impossibility of giving them adequate and satisfying opportunities. The few young shrews that do survive and actually acquire a place in the world are very lucky. They need food, water, a refuge above water from predatory fish, and below water or underground from predatory birds, a safe nest, and a surplus on which a wife can rear a family. Their dispositions become entirely changed: they are territorial, aggressive, conservative, just the opposite of their adolescent selves. One can see why: further questing will probably get them nowhere, and will almost certainly be fatal. This change, surely, is what one sees in other omnivores, including man. The adult male basically holds to what he has got, the adolescent is experimental, inquisitive and in revolt against him.

The last time the human species was ecologically really like any other was in the Palaeolithic. It was the Neolithic agricultural revolution that led (slowly) to our present state. But for western Europe one need go back only to the Middle Ages, and not that far for many other regions, to appreciate what strength of character was needed by individuals who could to some extent anticipate the future, if they were to persist and successfully rear a family. With epidemics raging unchecked, famine a constant threat and often a hideous reality, vast mortality of young children and of women in childbirth, and man's inhumanity to man to top it all, there could be little hope for any sort of reward in life except through inexperience (in adolescence) or a firm conviction of one's own importance so that one could impose one's requirements on others, or at least compensation could be expected in the next life for disasters in this. The human race has been selected not for objectivity of view but only to survive. It is true that in some practical things the two coincide. Either a particular berry is poisonous or not, and there is no going back on experience. In immediate practical matters, therefore, people are usually fairly objective. No doubt this is why technology has spread through the world: it produces indubitable practical results. But in all theoretical matters such as the long-term importance of the individual, those emotional attitudes will have survived that gave their possessors the strength to carry on, irrespective of any objective criteria. This is presumably why ideologies, unlike technology, are violently divisive. Obviously the exact nature of non-objective expression in each particular case will be largely due to cultural inheritance; I am only suggesting that a strong disposition to non-objectivity may have been selected for as well as strengthened by tradition.

The quotation above from Shaw will indicate clearly enough one emotional rejection of natural selection, and there are no doubt others, and many ways of rationalizing them. Now as population increases and the environment deteriorates, the necessity for taking as objective as possible a view of man, his nature, origin and relations to the rest of the world will become ever greater; yet more and more people will come into being who firmly reject an objective approach. There is a possible way out of the difficulty, possible because of the consequences of the Neolithic revolution. With vastly increased leisure, and facilities, especially medical, reducing the pressures of selection and aided perhaps by a considerable mingling of originally geographically separate populations, there is probably more human diversity in the world than ever before. Many more genotypes must now exist in some quantity than could persist in Palaeolithic and pre-Palaeolithic times, and there is a chance that some of them, deleterious then, are now of advantage in cultural evolution. In particular those with slow maturation, the perpetual adolescents, who retain their inquisitiveness, their desire to play, to find out what things are, may be of far more to the species than dominating and arrogant adults. It is likely that the university ecophene (at least in the biological sciences) is indeed composed largely of those who want to do research, although individuals as they age will tend to fall off into administration, and there may be some genetic characteristics of it. But if this is so, then this is in part an ecotype, selected into a particular niche, deserving careful investigation and far from representative of the human race. Presumably politicians are representative of the human race, and the contrast between them and research workers is only too apparent. Will the world in general with its vast numbers of rigid religious and political fanatics, its tyrants domestic, institutional, national and international, ever take serious notice of the tentative stumblings towards objectivity of evolutionists?

References (Cain)

Cain, A. J. 1951*a* So-called non-adaptive or neutral characters in evolution. *Nature, Lond.* **168**, 424.

Cain, A. J. 1951*b* Non-adaptive or neutral characters in evolution. *Nature, Lond.* **168**, 1049.

Cain, A. J. 1964 The perfection of animals. *Viewpoints Biol.* **3**, 36–63.

Cain, A. J. 1977 The efficacy of natural selection on wild populations. In *Changing scenes in natural sciences, 1776–1976* (ed. C. E. Goulden), pp. 111–133. Special publication 12, Academy of Natural Sciences, Philadelphia, Pennsylvania.

Lewontin, R. C. 1972 Testing the theory of natural selection. (Review of E. R. Creed (ed.) *Ecological genetics and evolution*. Oxford and Edinburgh: Blackwell). *Nature, Lond.* **236**, 181–182.

Lewontin, R. C. 1974 *The genetic basis of evolutionary change*. Columbia University Press.

[It has not been possible to include the general discussion in this publication.]